OTHER WORLDS

A VISUAL TOUR OF THE SOLAR SYSTEM

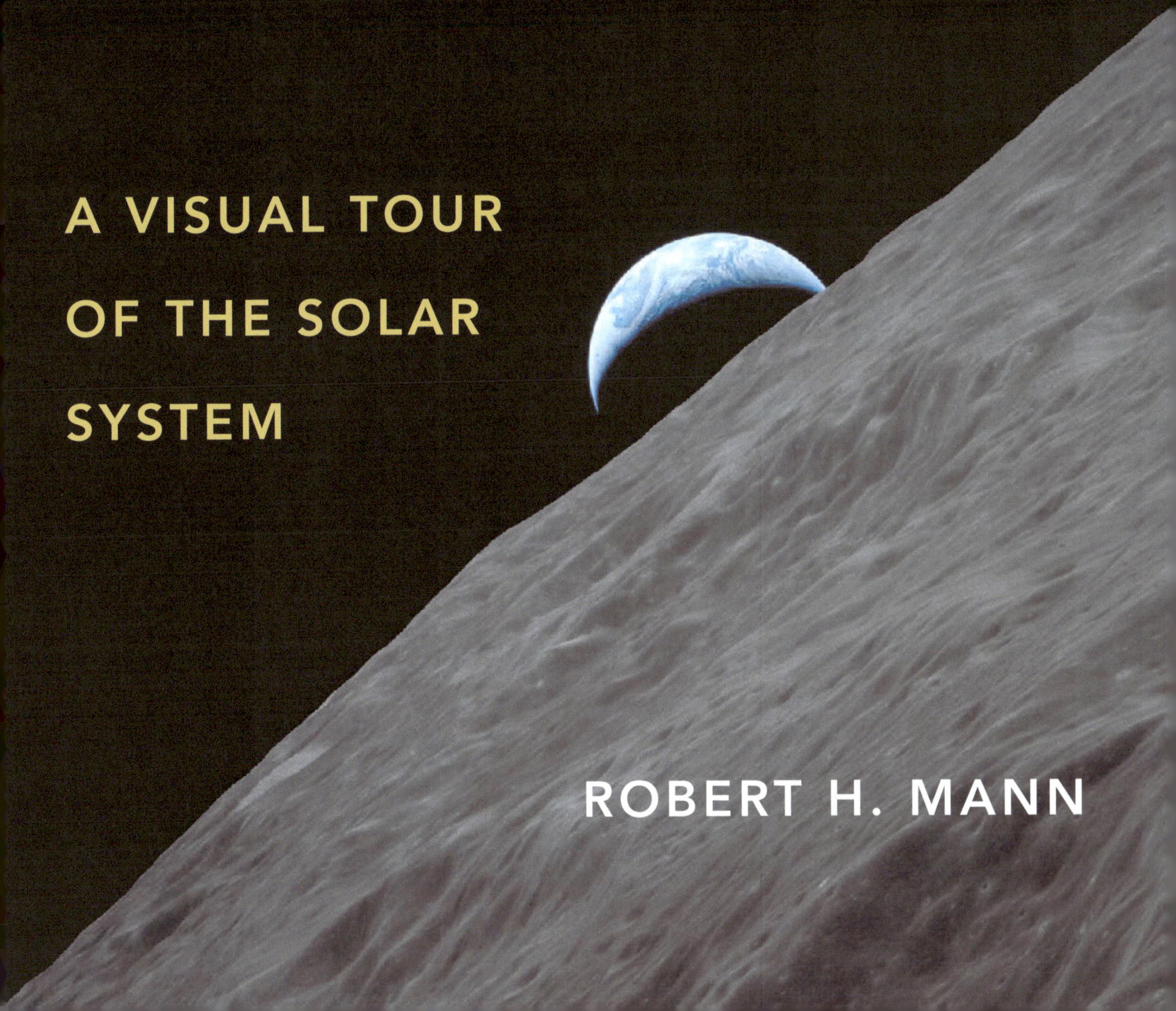

ROBERT H. MANN

For my wife and family

OTHER WORLDS: A VISUAL TOUR OF THE SOLAR SYSTEM

Published by Infinite Time Publications

For permission requests, please contact the publisher at the following email address:
infinitetimepublications@gmail.com

ISBN: 979-8-218-35016-1

Image credits: See pages 126-127

TABLE OF CONTENTS

INTRODUCTION

"Maybe the time is not quite yet. But those other worlds - promising untold opportunities - beckon. Silently, they orbit the Sun, waiting."

- Carl Sagan

WHEN PEOPLE LONG AGO looked up at the nighttime canopy of twinkling stars, they must have wondered about those few stars that drifted across the sky. Over a span of weeks and months, perhaps they noticed the slow-motion, graceful dances of these curious points of light. Ancient Greeks named them planetes, or wanderers. Today, we call them planets, and our wanderlust for them is stronger than ever.

For almost all of human history, we had little idea of the actual nature of these planets. Telescopes, first invented in the early 1600s, helped us see new things, such as the rings of Saturn, the shadows on the Moon, and the polar caps of Mars. It became apparent Earth was a planet itself, another majestic globe orbiting the central star we call the Sun. Planets, too small to burn inside or shine by their own light, were in fact not stars at all. But with details hidden, we were left to only speculate what tales these distant worlds could tell.

Since the 1960s, we have launched robotic spacecraft to fly by, orbit, and even land on the planets, all while gathering data and sending back magnificent pictures. In this short time, a blink of an eye in the 200,000 years humans have walked the Earth, the wanderers of our night sky have transformed into exciting, colorful, awe-inspiring *places*. The planets, moons, and other worlds orbiting the Sun are places we can imagine visiting, touching, and exploring.

Perhaps you, like myself, were born too early to experience these marvels in person. Yet, there is great joy in being alive in an age of discovery, an age when we are still learning about new worlds and can admire them in all their pristine beauty from afar.

In this tour, we will explore 20 of the most incredible worlds in the Solar System beyond Earth, which are grouped based on shared qualities. The first worlds are gaseous giants that helped sculpt the Solar System, while the next ones quietly tell us intriguing stories. Other destinations are dead but reveal beautifully chaotic features. The places thereafter are all active, thanks to tidal heating, and the final worlds are dynamic in both their skies and surfaces. We depart with a note on the wonders awaiting us. But first, we shall survey the Solar System as a whole, and take a closer look at its engine: the Sun.

Background photograph: Sunset on Mars, captured by NASA's *Curiosity* rover.

Artistic depiction of the birth of the Solar System.

ABOUT 4.5 BILLION YEARS AGO, a giant cloud of hydrogen gas and dust along the edge of the Milky Way galaxy began to contract under its own gravity. The particles pulled closer and closer into a hot, dense sphere. Eventually, the pressure was so intense that hydrogen atoms fused to create helium. This nuclear reaction generated a tremendous amount of energy, and the sphere began to shine. Our Sun was born.

The spinning Sun attracted and shaped the nearby gas and dust into a spiraling disk. Over time, dust within this disk began to stick together, forming small lumps of rock and ice called planetesimals. As some of the planetesimals grew, their gravitational force increased and they pulled in even more debris, becoming spherical planets. Shards of material in orbit around the planets followed the same process, forming their moons.

After a period of chaotic scrambling, in which some planets collided and others were ejected into deep space, things finally settled. Today, the eight planets stand out as the most dominant objects circling the Sun. The four closest to the Sun - Mercury, Venus, Earth, and Mars - are made of rock and are called terrestrial planets. The outer four - Jupiter, Saturn, Uranus, and Neptune - are much larger with thick, gaseous atmospheres and no solid surfaces. These are the giant planets.

Among the planets are the millions of smaller objects, or minor planets, leftover from the early Solar System. Some are asteroids, mostly found in the Asteroid Belt between the orbits of Mars and Jupiter. Beyond the orbit the Neptune, the Kuiper Belt is a disc of ice dwarfs. Another shell of icy debris called the Oort Cloud encases the Solar System. Occasionally, minor planets from these realms swing into the inner Solar System and become comets. Heated by sunlight, their surface ice vaporizes and forms a brilliant, long tail.

The center of all this action is the Sun, our closest star. From the biggest planet to the smallest asteroid, everything in the Solar System is gravitationally bound to the Sun. Without it, there would be no oceans, seasons, weather, life, or civilization on Earth. This star is so colossal that it makes up 99.8% of the total mass of the Solar System, and more than one million Earths could fit inside it. And yet, the Sun is an ordinary, medium-sized star, a mere grain of sand in the vast cosmic ocean.

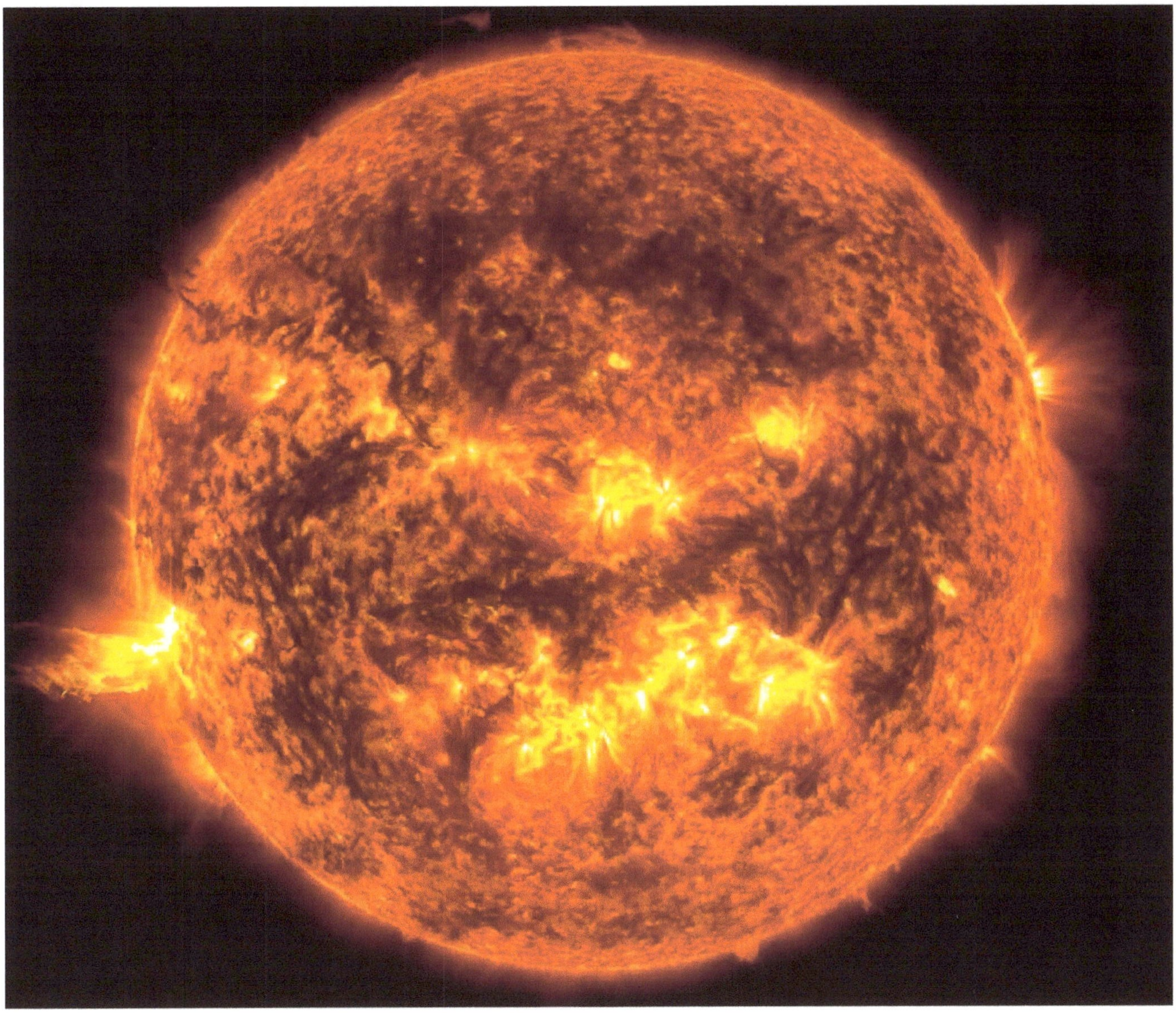

The Sun in ultraviolet light, from NASA's Solar Dynamic Observatory.

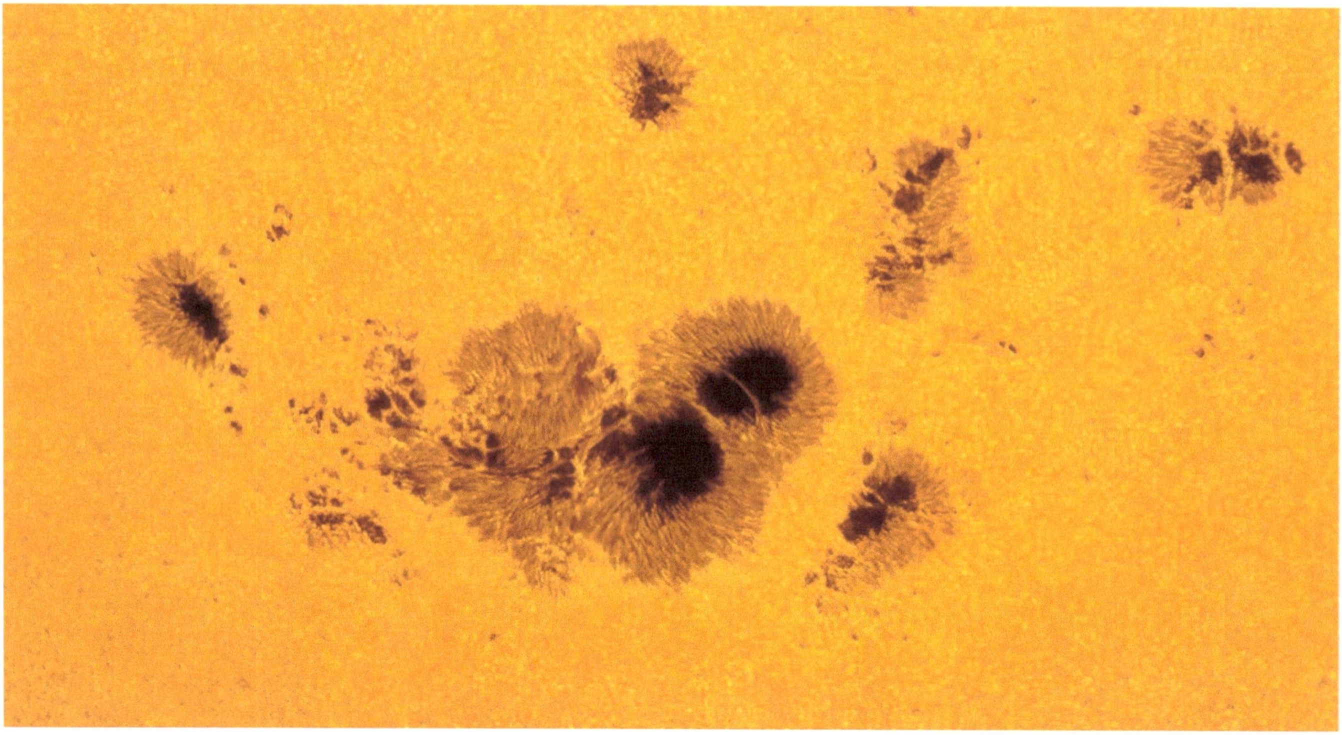

Several Earths could fit inside these largest sunspots seen from NASA's Solar Dynamic Observatory.

The Sun is made of super-hot, condensed gas called plasma. Hydrogen atoms are still fusing into helium in its interior, converting millions of tons of matter into energy every second, and powering the star's light like a giant nuclear bomb. Temperatures in this burning ball range from 10,000°F in the outer layer to 27,000,000°F in its core.

There is no solid surface on the Sun. The disk we see is where photons of light are able to finally break free from the churning interior and travel a straight line into space. From this disk, the light takes eight minutes to reach our eyes. But the photons within these rays were bouncing around inside the Sun for up to 30,000 years. The sunlight shining on us today was forged during the last ice age.

Our central star is the most powerful magnet in the Solar System, with its magnetic field generated by flowing gas in its interior. Sunspots are dark patches where strong magnetic currents disrupt its outer layers, growing and diminishing in an 11-year cycle.

Magnetic field lines expel hot gas in eruptions called coronal mass ejections. These blast billions of tons of charged particles into space, which interact with planetary magnetic fields and create beautiful aurorae. They can also disrupt our electronic systems, from cables to satellites to computers. Fortunately, a coronal mass ejection strong enough to upend our power grids has not happened since 1859, but that event reminds us our modern societies function at the mercy of the same star that gives us life.

The Sun, like every other star in the universe, cannot shine forever. As the Sun uses its fuel, it is slowly losing mass. Counterintuitively, this is causing it to slowly expand, since less mass results in weaker gravitational hold on its outer layers. In five billion years, the Sun will run out of hydrogen atoms to fuse and begin fusing helium into heavier elements. With this epic burst in energy, the Sun will swell into a red giant star, swallowing the innermost two planets, Mercury and Venus, and possibly even Earth.

When no more elements can be fused deep inside the Sun's interior, its energy will exhaust and the external layer will shed a colorful cloud of gas called a planetary nebula. After only a few tens of thousands of years, the gas will disperse, leaving behind the Sun as a shriveled white dwarf, a compact shell of its former youth. At this point, the Sun will no longer burn, but will continue to radiate low levels of heat for trillions of more years. If Earth is fortunate enough to survive the Sun's red giant phase and is still around to orbit the white dwarf Sun, it will spend the rest of its long life as a dark, barren, frozen wasteland.

Therefore, the Solar System as we know it today is a brief snapshot in time. But what a wonderful moment we can capture. It is time to journey to other worlds.

Artistic depiction of what Jupiter may look like orbiting the white dwarf Sun, tens of billions of years in the future.

Grand Sculptors

Giant Planet

Jupiter

WE BEGIN OUR TOUR with the largest and most massive planet in the Solar System. Jupiter, named after the Roman god of sky and thunder, casts a bright and rosy light in our night sky. Up close, Jupiter is a majestic sight to behold, with parallel bands of red and white clouds, cuffed with wispy streaks. The white zonal clouds are likely made of ammonia ice, whereas the composition of the red belts is unknown. In the polar regions, blue swirls tango around as if they were in a cosmic Van Gogh painting. Befitting its name, powerful lightning bolts frequently flash from these icy tapestries.

The most stunning feature of Jupiter is its Great Red Spot. Winds along the edge of this spiral of gas reach up to 245 mph, and it is so large it could engulf two Earths. Often compared to a hurricane, the Great Red Spot is actually a high-pressure system, spinning counterclockwise in Jupiter's southern hemisphere. It is also remarkably stable, having persisted for at least 150 years with no change in position.

To answer how such a storm could rage for so long, we must consider the bulk properties of the planet. Unlike Earth, Jupiter has no solid surface, belonging to a class of planets known as gas giants. With no solid ground to create friction, circulations in the atmosphere can last uninterrupted for a very long time.

As a gas giant, Jupiter is primarily composed of hydrogen and helium, the lightest elements in the universe, overlaying a relatively small core of rock. But because the planet is so massive, its extreme gravity pressure-cooks the hydrogen deep inside its interior so that it is liquified. The transition from the gaseous to liquid layer is gradual, perhaps with droplets of helium and neon raining down into a dense, hydrogen ocean. This ocean is certainly unlike any we know of, with a pressure more than three million times that of the surface of Earth, and the "water" of hydrogen acting more like a molten metal.

Jupiter's Great Red Spot, photographed by NASA's *Juno* orbiter.

Such a bizarre internal structure explains another dramatic feature: its magnetic field. As the planet rotates, the liquid hydrogen within it generates electric currents that fuel the most powerful planetary magnetosphere in the Solar System. As on Earth, the interactions between Jupiter's magnetosphere and the solar wind create aurorae at the poles, but Jupiter's neon blue lights are much more spectacular than those on Earth.

Jupiter's magnetosphere is also huge. If it were visible to the naked eye, it would appear in the sky in the shape of a teardrop and larger than the full Moon. Strong magnetic currents generate radio emissions, which can be heard from radios on Earth at frequencies of roughly 10 to 40 Mhz.

Jupiter has played a major role in sculpting the Solar System. The belt of asteroids just inside its orbit would have formed into a planet, were it not for Jupiter's gravitational interference. Early migrations in the giant planet's orbit may have bombarded Earth with asteroids and comets that delivered the water needed for life. Asteroids and comets caught in Jupiter's gravitational net were either flung out of the Solar System, pulled into Jupiter itself, or thrown into Earth's neighborhood. One asteroid, likely steered off course by Jupiter, crashed into Earth 65 million years ago, killing the dinosaurs and setting the stage for the evolution of humans. Other small objects which could have hit Earth were instead deflected elsewhere by this giant planet. Jupiter thus emerges as a creator and destroyer of life, an architect of the entire story of humankind and of our very existence.

Enhanced real photograph of storms swirling in the polar area of Jupiter, from NASA's *Juno* obiter.

Jupiter as seen from NASA's *Cassini* spacecraft.

Saturn

AMONG ALL THE PLANETS visible to the naked eye, the slowest moving is Saturn, beaming in the night sky as a pale yellow point of light. Recognized for its slow and dignified motion in the heavens, Saturn was named after the father of Jupiter, and the Roman god of wealth and agriculture. Seeing Saturn in all its glory, we realize we have stumbled upon the Solar System's crown jewel, a dazzling cosmic creation and a stir to the imagination.

Like Jupiter, Saturn is a gas giant, composed primarily of hydrogen and helium, with no solid surface. Although the second biggest planet in the Solar System, Saturn is the least dense. In fact, if the planet could somehow be dropped into a larger tub of water, it would float.

Deep in the interior of this beach ball planet, the gaseous clouds gradually transition to a soupier, condensed liquid, possibly with helium and neon droplets raining into an ocean of metallic hydrogen. All of this surrounds a small rocky core. The liquid in Saturn's interior churns and generates electric currents, producing its magnetosphere. The interaction between Saturn's magnetosphere and the solar wind creates the bright blue aurorae, draped around its poles.

Due to mixing of chemicals, Saturn's upper atmosphere is tinged in shades of gold. Cloud jets whip across the planet at staggering speeds of up to 1,100 mph. Unlike Jupiter, Saturn's axis is tilted by about 27 degrees, giving it seasons. One unique seasonal tradition on Saturn is the Great White Spot, which has been observed from telescopes since 1876. As spring begins in the northern hemisphere, and perhaps due to accompanying changes in temperatures, small white thunderstorms form and rapidly expand horizontally, sometimes encircling the entire planet, before eventually dissipating.

Saturn also hosts a bizarre, permanent polar vortex in the shape of a hexagon. Large enough to fit Earth, this great hexagon of the north pole is kept in place due to complex atmospheric interactions and is unlike anything seen elsewhere in the Solar System.

The polar hexagon of Saturn, imaged by NASA's *Cassini* orbiter.

When astronomer Galileo Galilei first observed Saturn through his telescope in 1610, he found what appeared to be two ears poking out from its sides. Two years later, these ears disappeared, and reappeared again in 1613. Confused, Galileo wrote: *"I do not know what to say in a case so surprising, so unlooked for and so novel."*

Galileo had unknowingly discovered one of the most iconic features in the universe: Saturn's rings. As Saturn wobbles along its axis, the angle of the rings changes as seen from Earth. The reason why Galileo reported them as missing is because at that time the rings were in the same plane as Earth, and so were edge-on and too flat to be seen.

Saturn's rings are 175,000 miles wide, which is almost the distance between the Earth and Moon. Viewing Saturn from afar, its rings appear as a glamorous, glistening grooved disk enveloping the golden planet. But the rings are actually made of billions of water ice flakes, ranging in size from smaller than a grain of sand to the size of a car. Distinct gaps in the rings are caused by small moons that sweep up debris and help keep the ring chunks in place, like shepherds tending their sheep.

Most of Saturn's rings are only 30 feet thick. But icy particles at the edge of one of the rings tower in the shape of mountains, miles above the plane, casting shadows upon the main rings below. These breathtaking ring mountains are likely caused by gravitational tugs from nearby small moons. Other small moons bend the ring material as they orbit Saturn, creating a series of ripple waves along the edge of the rings.

Peaks casting shadows on the flat plane of Saturn's rings, imaged by NASA's *Cassini* orbiter.

NASA's *Cassini* orbiter catches a small shepherd moon, Daphnis, making waves in Saturn's rings

Sunlight shining on Saturn and its rings, imaged by NASA's *Cassini* orbiter.

Earth, spotted from Saturn by NASA's *Cassini* orbiter.

Just as a new carpet collects dust, so too should ice in planetary rings. Because Saturn's rings are pristine, they are believed to have formed relatively recently, perhaps less than 400 million years ago. While this may seem ancient on a human timescale, Saturn itself is about 4.5 billion years old.

The origin of the rings is uncertain, but may have involved the shattering of a former moon. Studies suggest the rings will disappear in 100 million years, so their presence just as humans could see and explore them may be a fortunate coincidence.

Saturn is not a creator and destroyer of life, as is Jupiter. But early in the Solar System, Jupiter probably began migrating inward. In many other solar systems, gas giants are found orbiting close to their parent star, so our largest planet could have easily worked its way to a similar orbit, consuming Earth and ending any hope of civilization evolving in the Solar System.

What stopped Jupiter in its tracks and reeled it back in was Saturn, a planet massive enough to gravitationally dance with Jupiter. The end result was Jupiter settling in its current position, and Earth continuing unharmed, free from bullies. Poetically, Saturn, the father of Jupiter and the god of wealth and agriculture in Roman mythology, may have disciplined its wayward giant son and allowed the bountiful flourishing of life on Earth.

Giant Planet

Uranus

THE NEXT PLANET on our tour, Uranus, is so far away that it is only visible to the naked eye in the darkest of nights. When astronomer William Herschel discovered Uranus in 1781, he initially thought it was a comet, since at that point no new planet had ever been recognized. Named for the Greek god of the sky, the skies of Uranus may be the most mysterious and misunderstood of all the planets.

When Uranus was first seen up close in 1986, it presented a bland, featureless face, a stark contrast to the complex storm patterns seen on Jupiter and Saturn. However, telescopes from Earth have revealed a dynamic planet awoken from its slumber, with dark swirling spots popping up and disappearing, lightning storms ablaze, and a bright white vortex capping its pole.

The key to understanding this change in activity is in Uranus's axial tilt of 98 degrees. The planet is tipped on its side, rolling around the Sun like a barrel. The cause of this tilt is uncertain, but it is suspected that Uranus suffered an ancient collision with another large planet.

The tilt results in the most extreme seasons in the Solar System. Summer on Uranus is 21 years of continuous sunlight, and winter is 21 years of complete darkness. During equinox, when its equator is pointed toward the Sun and the hemispheres receive equal amounts of sunlight, the changes in temperatures stir up the planet's atmosphere, giving it new life.

As Uranus tilts, so do its moons and rings, all orbiting the planet in the same up-to-down fashion. Uranus's rings are much thinner and darker than Saturn's, reflecting about as much sunlight as coal. The dusty flakes within the rings range in size from the size of a pebble to the size of a house. Shepherd moons sculpt the rings, as they do for Saturn's rings. Curiously, one of Uranus's rings appears to breathe, expanding and contracting by three miles every several hours, for reasons currently unknown.

Picture of Uranus's chalky rings from NASA's *Voyager 2* spacecraft.

Like Jupiter and Saturn, Uranus is a giant planet with no solid surface. Its atmosphere is made mostly of hydrogen and helium, with traces of other hydrocarbons such as methane, which imbue it with a hazy turquoise color. Deep inside, the atmosphere transitions to a hot and extremely dense layer of water and ammonia. This water-ammonia ocean overlays a small rocky core.

Scientific experiments hint that the methane in Uranus's atmosphere breaks down into separate carbon and hydrogen atoms, with the carbon atoms forming diamond hailstones up to millions of carats in weight. These giant diamonds then fall through the mantle. Because of this odd internal structure and abundance of volatiles which easily freeze, including exotic forms of water, Uranus belongs to a class of planets called ice giants.

The ocean of high-pressure water and ammonia is an excellent conductor of electricity, so it is no surprise that Uranus hosts a magnetic field. However, Uranus's magnetic field is shifted a few thousand miles from the center and is tilted 60 degrees from the planet's rotational axis. The cause of these peculiarities is a mystery, although it may be that the field is generated at a more shallow depth than on other types of planets. Bright patches of aurorae form over Uranus's polar regions, resulting from the solar wind meeting this unusual magnetosphere.

NASA's *Voyager 2* spacecraft imaged Uranus's rings in false color to help study their nature and origin.

Uranus photographed in near-infrared light by the *James Webb Space Telescope*.

Uranus, as far as we can tell, has not as profoundly impacted life on Earth as have Jupiter and Saturn. Its formation surely disrupted smaller objects nearby, several of which may have bombarded early Earth. But the story of Uranus, at times melancholy and at times vigorous, is not complete. Its discovery was a scientific sensation, revolutionizing our understanding of the Solar System and spurring searches for even more planets. As recent observations show, we have still barely begun to understand this lopsided, turquoise gem. It beckons us.

Giant Planet

Neptune

WHILE URANUS WAS DISCOVERED by accident, it took years of toil and a lot of math to find Neptune in 1846. Named after the Roman god of the sea, it is the farthest planet from the Sun and is too distant to see without a telescope. Up close, Neptune is a pale blue giant, with dark spots as large as Earth, and white feathery clouds gliding merrily along.

Right: Neptune's high-level clouds as seen from NASA's *Voyager 2* spacecraft.

Neptune's dark spots are vortices which last for up to a few months. These spots appear to be at lower altitudes than the surrounding atmosphere, but what causes their formation, movement, and disappearance is not well understood. Since Neptune's axis is tilted about 28 degrees, it experiences seasons, so perhaps these storms are driven by changes in sunlight as the seasons progress. Also mysterious is the fact the farthest planet from the Sun has the strongest winds in the Solar System. Winds on Neptune blow up to 1,300 mph, faster than the speed of sound on Earth.

Neptune is an ice giant like Uranus, a planet made with a significant amount of volatiles such as strange forms of water. The atmosphere is composed of mostly hydrogen and helium gas, with some methane and hydrocarbons which reflect blue light, giving Neptune its color.

Underneath the atmosphere is a thick layer of fluids containing super-dense water and ammonia, which surrounds a small rocky core. Studies suggest that the carbon atoms in Neptune's atmospheric methane are separated to form diamond crystals that weigh up to millions of carats and rain down through the mantle, just as what may occur in Uranus.

The liquids sloshing in this bizarre interior generate electricity, fueling Neptune's magnetic field. The magnetic field of Neptune is offset from the center by several thousand miles and strongly tilted relative to its rotational axis, similar to the one seen at Uranus. That both ice giants have such misaligned magnetic fields indicates that these structures are driven by shallower fluids. Despite these similarities, the aurorae on Neptune are rather weak compared to those seen on Uranus.

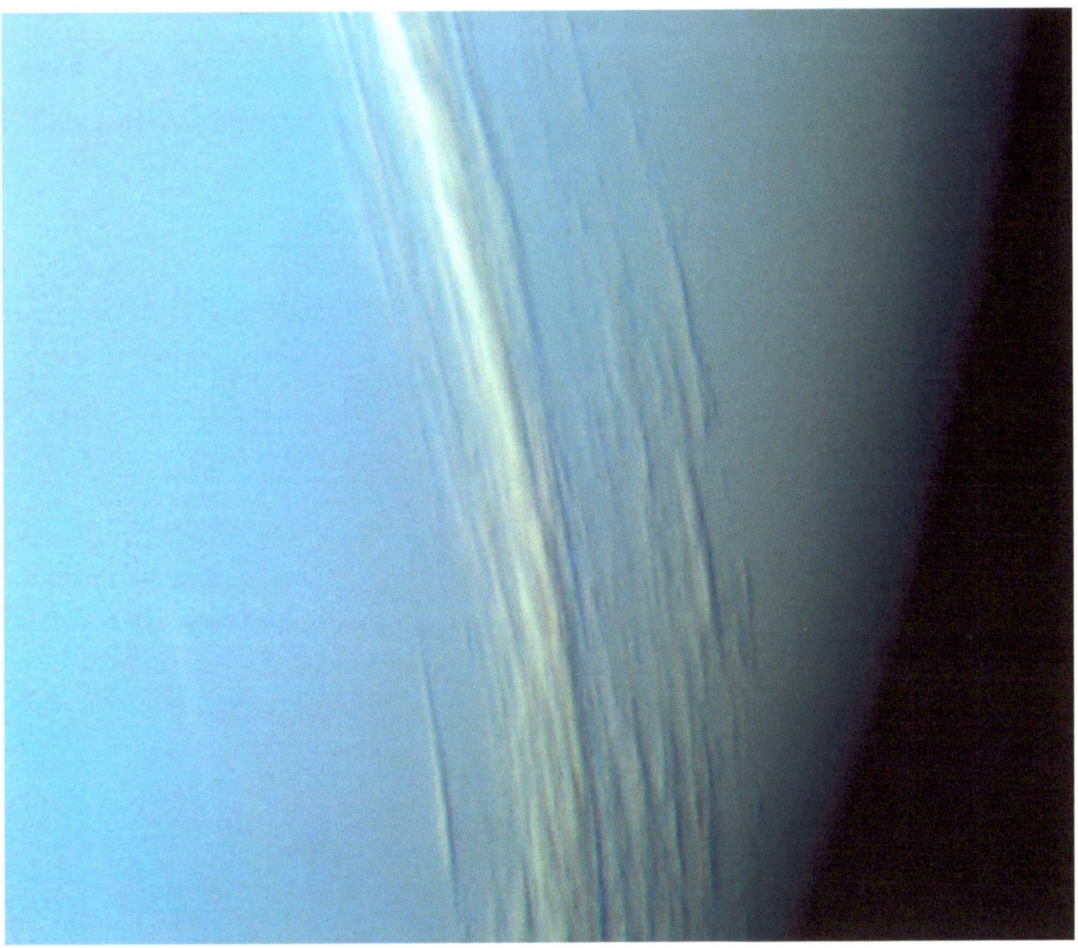

NASA's *Voyager 2 spacecraft* spots the bright arcs within Neptune's faint rings.

Neptune also hosts rings, although they are fewer in number and less complex than the rings of Uranus. Neptune's rings are extremely dark, reflecting only a small amount of sunlight, and contain a significant amount of dust. Neptune's outermost ring is comprised of five arcs, or sections, that appear much brighter than the rest of the ring.

If we could view this ring in fast motion from above, we would see the faint, seemingly stationary line of the ring with the arcs whizzing by in orbit, like the blur of a high-speed train rounding the curve of a railroad. These arcs may be maintained by complicated orbital interactions with small moons of Neptune.

The formation of Neptune in the early Solar System probably triggered an onslaught of small objects into Earth's neighborhood, with some impacting Earth itself. Today, Neptune's main role in the Solar System is as the lord of the Kuiper Belt, a diffuse disk of icy debris beyond its orbit. Many Kuiper Belt objects are captured by Neptune into orbital resonances, in which they circle the Sun in perfect rhythm with the planet. These configurations help keep the Kuiper Belt ice dwarfs stable. Others are free from such shackles, and are at greater risk of getting flung inward, to become comets and tangle with Jupiter.

If Jupiter is the creator and destroyer of life, Neptune is the gatekeeper, the distant blue marble deciding which objects to let in and wreak havoc, and which to keep safely afar.

The dynamic and ever-changing atmosphere of Neptune, captured by NASA's *Voyager 2* spacecraft.

Stories to Tell

Terrestrial Planet

Mercury

OUR JOURNEY NOW SWINGS US from the distant reaches of the Solar System to the closest planet to the Sun. Mercury, named for the speedy Roman messenger god, only appears from Earth as a small, bright point of light near the Sun as it sets and rises. The planet indeed moves quickly, circling the Sun every 88 days. Unlike the four giant planets we have toured, Mercury is an airless rocky planet, a scarred solid face bearing witness to all its past turmoil.

With no atmosphere blanketing the planet, Mercury experiences the most extreme temperature fluctuations in the Solar System. Daytime temperatures on Mercury reach a scorching 800°F, hot enough to melt lead, while nighttime temperatures plunge to -290°F. Paradoxically, this innermost planet has deposits of water ice, hundreds of feet thick, trapped inside craters which are permanently shadowed from the Sun. This ice was likely delivered to Mercury by comets impacting its surface.

Mercury has the highest uncompressed density of any planet in the Solar System, indicating it is composed significantly of the heavy element iron. This small planet likely has a solid inner core of iron, a molten iron outer core, and a thin mantle and crust of silicate rock.

Mercury's higher abundance of iron compared to other planets is not understood. The most obvious theory is that Mercury was struck by another object early in its history, stripping most of its outer silicate crust. However, subsequent observations reveal the planet has a high concentration of lighter elements such as potassium and sulfur on its surface, which would have been vaporized by such a collision. So, Mercury does not give up its secrets that easily.

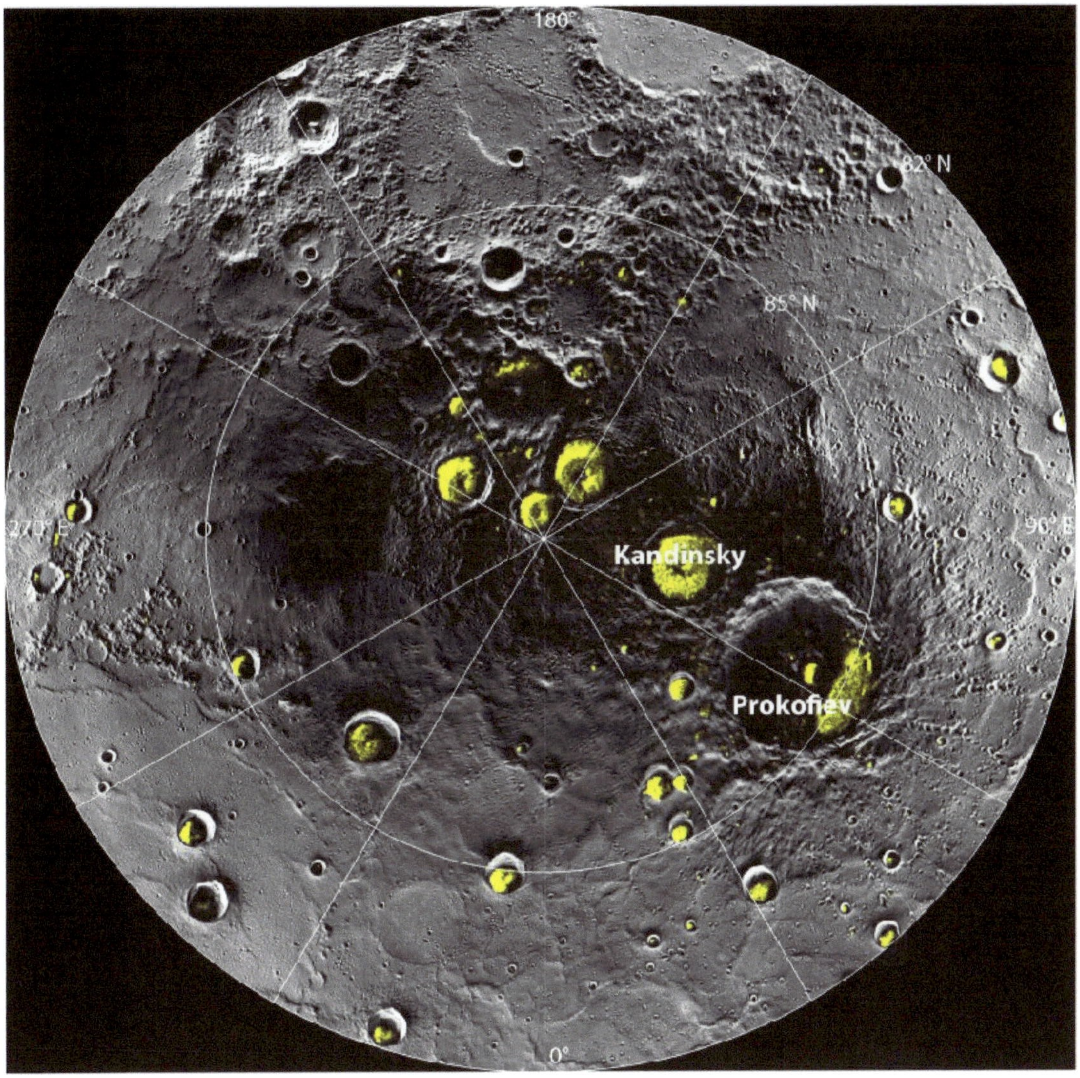

Deposits of water ice, shown in yellow, in Mercury's north polar region.

The molten iron inside generates a weak magnetic field, making Mercury the only inner terrestrial planet besides Earth to do so. Its magnetosphere interacts with the solar wind, as do other planetary magnetospheres. But because Mercury has no atmosphere, the charged particles from these interactions actually produce X-ray aurorae on the surface. On Earth, we can look up at the sky and marvel at the northern lights. If we could stand on Mercury and wear X-ray goggles, perhaps we might look down and see the northern lights glowing in the ground beneath our feet.

Mercury's surface is covered in craters, created from the many impacts it withstood from colliding asteroids and comets. Because new processes would erase craters, their prevalence indicates Mercury's surface is old and has changed little in the past three billion years. But volcanism clearly played a role early in its history, with vast smooth plains having once been covered by extensive lava flows.

Volcanic plains of Mercury, photographed by NASA's *Messenger* orbiter and processed to represent approximate natural color.

View of Mercury from NASA's *Messenger* orbiter.

Mercury also hosts impressive cliffs, some hundreds of miles long and up to two miles high. Just as a grape shrivels into a raisin, these cliffs are wrinkles formed as Mercury has slowly cooled and shrunk. Some smaller cliffs appear rather fresh, indicating this process is still ongoing. Other recent, localized activity involves the creation of hollows, small clusters of holes and halos found inside craters. These are thought to form as light elements on its surface sublimate, or turn from solid to gas, due to sunlight exposure.

Mercury may be heavily battered, but it is not completely dead. We have not yet read all the pages of this world's story.

The Moon

FOUR AND A HALF BILLION YEARS AGO, the Solar System was a crowded and messy place, with planets and other smaller objects frequently colliding and competing for dominance. During this chaotic period, a young Earth found itself in the crosshairs of another planet. The smaller planet crashed into Earth, disintegrating and unleashing a swarm of fragmented rocks and dust into space. This debris was then caught in orbit around Earth and began clumping together. In just a short time, the Moon was born.

The Moon, made of ashes blasted off Earth, has been a stirring sight in the night sky for as long as wandering eyes have gazed upon its light. Suspended among the quilt of stars like an ethereal orb, no other world beyond our own has had such a profound bond with humanity.

For most of history, the Moon was the only nighttime source of light, allowing for navigation and hunting. The phases of the Moon as it orbits Earth relative to the Sun, alternating between new, crescent, full, crescent, and back to new in 27 days, gives us our concept of months. Because the Moon completes a full rotation in the same time it takes to orbit the Earth, one side always faces us, with the other side forever hidden from our earthly view. The Moon we see at night is the exact same face our ancestors saw when they looked up 200,000 years ago.

On rare occasions, the Moon perfectly aligns with the Sun and block most of its light. These stunning cosmic events, solar eclipses, are unlike any seen elsewhere in the Solar System. Their fact and nature owe much to random coincidence. If the Moon were a little smaller or farther away, it would not be large enough in the sky to cover the Sun. If the Moon were a little larger or closer, it would be big enough to hide the Sun's outer corona, robbing us of the iconic "black hole sun" appearance of eclipses.

From the Moon, Earth is indeed large enough that when it eclipses the Sun, the solar corona is blocked. However, Earth's own atmosphere is backlit by the Sun, surrounding it with a brilliant orange ring instead. Future inhabitants of the Moon will surely marvel at these unique, ring-of-fire eclipses.

Artistic depiction of a solar eclipse as seen from the Moon.

Sunrise over the central mountain of Tycho Crater, photographed by NASA's *Lunar Reconnaissance Orbiter*.

Like Mercury, the Moon lacks an atmosphere, and so temperatures swing from 250°F during the day to below -200°F at night. The Moon also resembles Mercury in its surface. Most of the Moon is covered in craters, scars from ancient impacts by smaller objects. Its large darker areas, named the maria after the Latin word for seas, are actually plains covered in lava flows that solidified more than three billion years ago. Interestingly, the side of the Moon that never faces Earth has very few of these volcanic plains, indicating the Moon's crust is thinner on the side facing Earth. Below this uneven crust is a mantle of silicate rock, overlaying a small inner core of iron.

Life on Earth would be different without the Moon. As it orbits, its gravity tugs on Earth's oceans, causing tides. Tide pools play a significant role in coastal ecosystems, and may have even been essential in the evolution of creatures from the sea to land. The Moon also stabilizes Earth's axis. As Earth spins like a top, it sometimes wobbles. Such wobbles are historically associated with changes in climate, such as ice ages. Without the Moon acting as an anchor, the wobbles would be greater, and Earth may well have experienced climate swings too catastrophic for humans to survive.

But as Earth's rotation is very gently slowing down, so too is its gravitational hold on the Moon, which moves 1.5 inches away each year. The story of the Earth and Moon will end in one of two ways: either the Sun will engulf both in five billion years when it balloons into a red giant star, or the Earth and Moon will survive this challenge and the Moon will free itself from Earth's orbit in 50 billion years. One way or another, this cosmic love affair is fated to die.

The Earth and Moon, two floating marbles in the vastness of space, captured by NASA's *Galileo* spacecraft.

Asteroid

Ceres

IN 1801, ASTRONOMER GIUSEPPE PIAZZI FOUND a new planet between Mars and Jupiter and named it Ceres, after the Roman goddess of agriculture and fertility. More planets were soon found in similar orbits, and all appeared to be much smaller than the other known planets. Eventually, these objects were reclassified as asteroids, and are part of what is now recognized as the Asteroid Belt. But Ceres, the largest and most massive asteroid, is the only one which is clearly round due to gravity, like the planets. The surface of Ceres appears bland, but befitting its original classification, there is more than meets the eye.

Like Mercury and the Moon, Ceres is a solid object with no atmosphere. Unlike those two worlds, and also unlike many asteroids, Ceres's lower density indicates it contains significant amounts of both rock and ice. It likely has a core of silicate rock, surrounded by a mantle of water ice and a thin, dusty outer crust. Temperatures on Ceres range from -100°F in the daytime to -225°F at night.

Towering from the heavily cratered plains is one lonely mountain 2.5 miles high, called Ahuna Mons. This dome is well formed, with bright streaks of sodium carbonate draping its slopes, suggesting it is perhaps no more than 200 million years old. Ahuna Mons is possibly a volcano that erupted a muddy slurry of rock and ice, a mixture unique in the Solar System. Oddly, Ceres has no other obvious volcanic structures. Its volcanoes may flatten within a few hundred million years, due to the bedrock ice being mobile enough at the asteroid's temperatures.

Ceres is the darkest of all the worlds we will see on this tour, reflecting less than 10% of sunlight, comparable to freshly paved asphalt. But a cluster of bright white spots, beaming from the dark lands like searchlights in the night, are located inside a crater named Occator. A closer look in the crater reveals the spots are deposits of salts mixed with ammonia, the residue of briny water eruptions from its interior. The eruptions probably began only 20 million years ago, and brine may still be percolating up to the surface. Even more interestingly, it is suspected that hydrothermal activity deep underground drove these eruptions after the crater-forming impact. On Earth, such warm, water-rock interactions occur in seafloor hydrothermal vents, where life thrives today, and where life on our planet may trace its origin.

View of Ceres from NASA's *Dawn* orbiter, showing the bright spots inside Occator Crater.

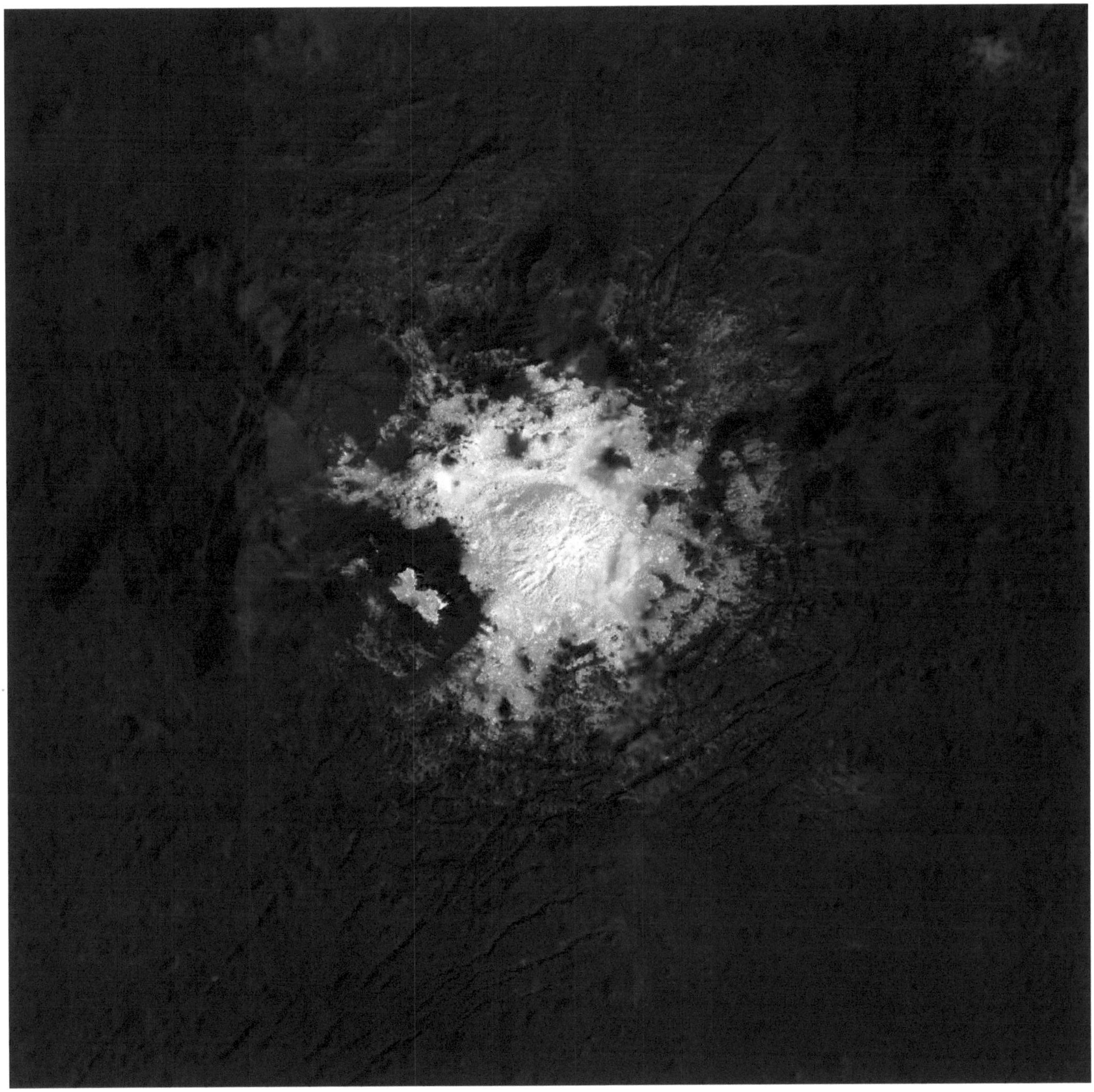

Closeup of the salt deposits in Occator Crater, as seen by NASA's *Dawn* orbiter.

The bright spots invite us to dig deeper into Ceres's history. Ammonia ice in these deposits hints that Ceres did not form in the Asteroid Belt, but instead originated farther out, where ammonia is more abundant. Where exactly this asteroid formed and how it journeyed into its current position is not yet established. Ceres's interior was heated by radioactive decay in its core, which likely melted the icy mantle into a liquid water ocean. While this ocean should now be mostly frozen, pockets of watery brines seem to still exist roughly 25 miles below the surface, as evidenced by its bright salt deposits.

Questions regarding the origin of Ceres, how much liquid water still exists in its interior, and whether or not it ever sustained conditions suitable for microbial life to emerge, will keep us pondering this diamond in the rough.

Moon of Jupiter

Callisto

GALILEO GALILEI, THE SAME ASTRONOMER who first observed Saturn's rings, made an even more important discovery in 1610. Pointing his telescope at Jupiter, he found four moons in orbit around the planet, thus proving not all heavenly objects revolved around Earth. These Galilean moons ignited a scientific revolution and are incredible worlds in their own right. Our journey now brings us to the one farthest from Jupiter, and the most serene of them all.

Callisto, named after a female lover of Zeus in Greek mythology, is a dark moon sprinkled with bright, sparkly craters as if it were a cosmic disco ball. The smooth, dark material on its surface is probably composed of silicate rocks mixed with carbonaceous compounds. Icy spires several hundred feet tall jut out from this landscape. No other world in the Solar System has these kinds of features. Callisto's ice peaks likely form when water ice along crater rims slowly sublimates, changing from a solid to a gas. The residual dusty debris descends the slopes, leaving behind the tall spires of ice, which themselves erode over time.

Some of Callisto's craters are truly a sight to behold. Valhalla, the largest multi-ring structure in the Solar System, is a majestic bullseye 2,360 miles in diameter, nearly as wide as the continental United States. The concentric rings surrounding the central crater of Valhalla probably formed when the impact punctured a liquid or soft icy layer below the harder crust. This supports the idea that Callisto had a subsurface water ocean, at least early in its history.

Icy spires of Callisto, captured by NASA's *Galileo* orbiter.

Craters and bumps dot Callisto' surface, as captured by NASA's *Galileo* orbiter.

If Callisto still has an ocean, it is buried deep under the crust, and sits atop high-pressure ice. Callisto may not even have a fully separate core of rock, but rather a mix of rock and ice all the way to the center. Therefore, any ocean inside this moon, not interacting with the surface nor a hot, rocky interior, would be lifeless and sterile.

And yet, if humans are destined to settle the outer Solar System, Callisto will probably be the first world beyond Mars we call home. The other large moons of Jupiter are tucked within Jupiter's deadly radiation belts, which will complicate any human missions to those worlds. On the other hand, Callisto orbits safely outside of this danger zone, hosts no violent eruptions or storms to worry about, has a surface gravity more familiar to us than any of the asteroids, and is full of useful resources such as water ice. Perhaps our future descendants, stepping foot outside their glass dome and gazing up at Jupiter hanging in the night sky like a hot air balloon, will be deeply moved by Callisto's tranquil beauty, and the elegant interplay of ices on its surface that sculpt its jagged peaks.

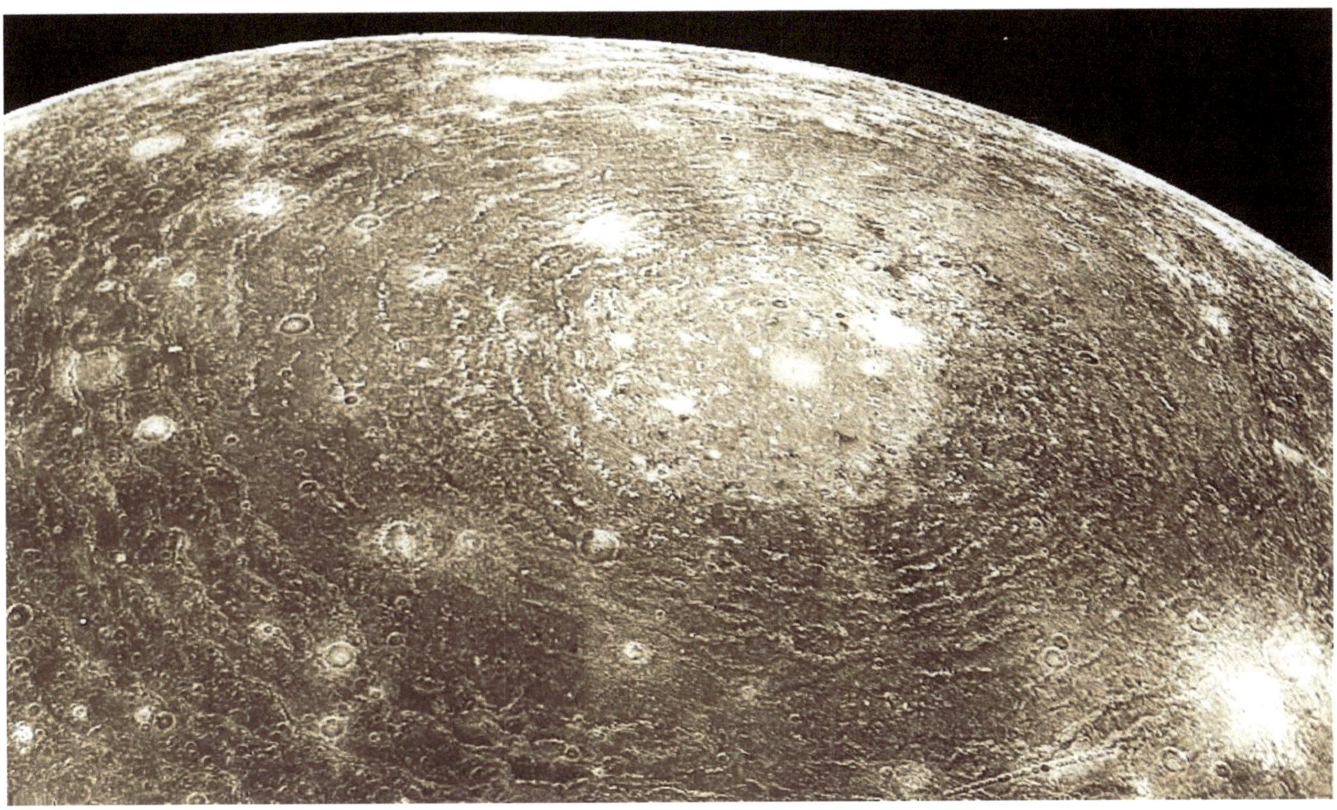

Valhalla, a continent-sized, multi-ring impact structure, imaged by NASA's *Voyager 1* spacecraft.

Hints of Chaos

Moon of Jupiter

Ganymede

BEHOLD GANYMEDE, the largest moon in the Solar System and bigger than planet Mercury. Named for the male lover of Zeus in Greek mythology, Ganymede is one of the four moons of Jupiter astronomer Galileo Galilei discovered in 1610. In contrast to the simplicity of its neighbor Callisto, Ganymede boasts a magnificent geological history and more liquid water than on Earth or anywhere else in the Solar System.

Ganymede is a world paved with grooves, crisscrossing an otherwise dark and cratered surface like a complex superhighway system. These grooves are relatively light features, several tens of miles wide and hundreds of miles long. They are also not as heavily cratered as the darker terrain and so must be younger, perhaps two billion years old. Most likely, Ganymede's crust stretched as it expanded early in its history, splitting its surface and causing the wide and shallow swath of grooves we see today.

Why the grooves are lighter than the surrounding terrain is still an open question. Either the dark terrain was pulled apart and brighter ice underneath was exposed, or icy lava seeped from below and flooded the grooves. Several small, scalloped depressions located throughout the grooved terrain could be sites where this lava erupted, hinting that volcanism, at least on small scales, has played some role in shaping these grooves. If lava flowed on Ganymede's surface several billion years ago, it would have been composed of slushy water mixed with other chemicals such as ammonia.

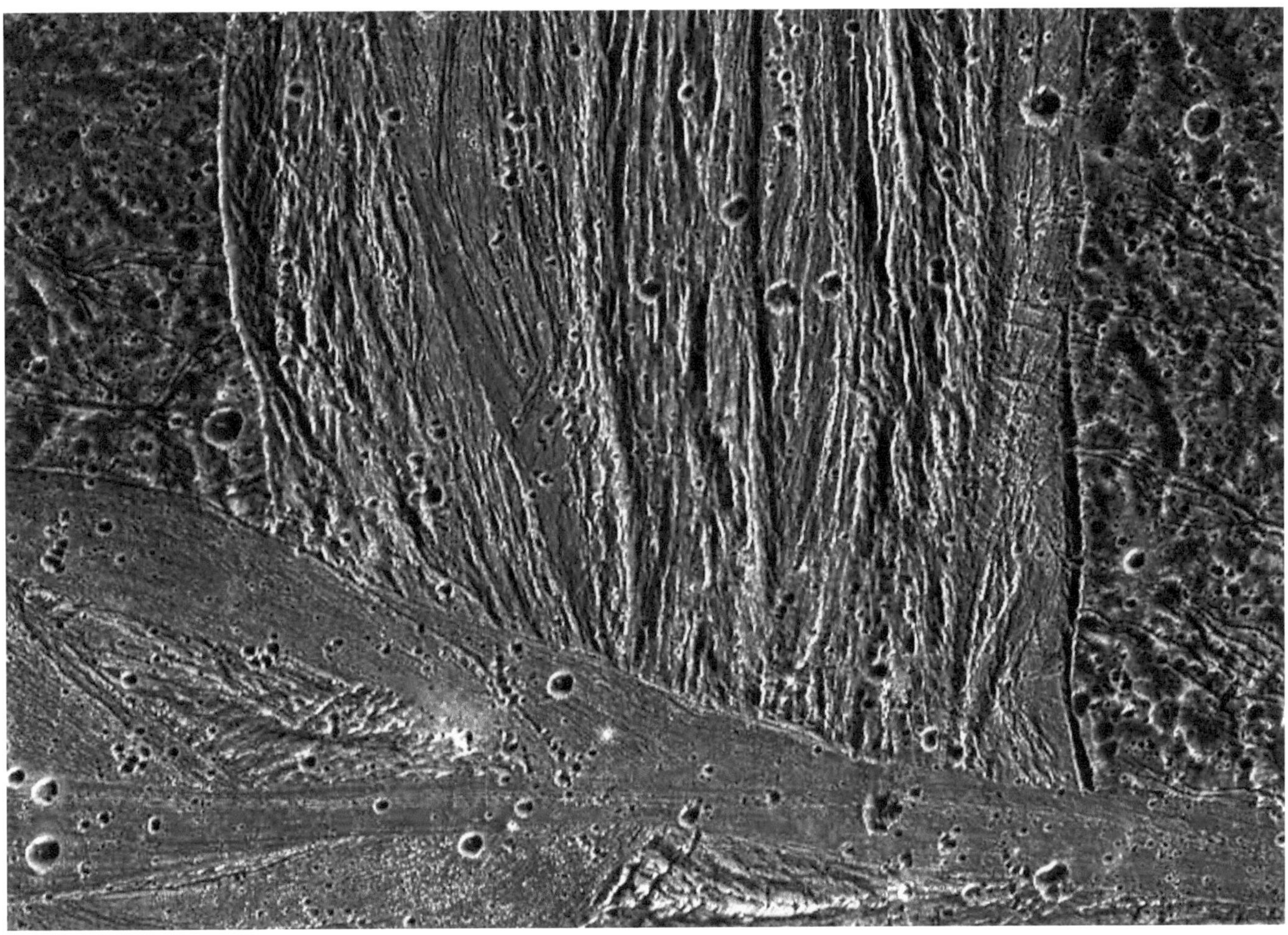

NASA's *Galileo* orbiter captures an intersection of Ganymede's grooves.

Ganymede has other unique surface features. Parallel ridges, or furrows, extend across the dark terrain, up to hundreds of miles long and spaced tens of miles apart. In one dark region with furrows covering roughly the size of Australia, it appears as if a cosmic farmer plowed the entire continent-sized land. Ganymede's furrows may be the aftershocks of a massive, ancient impact. Ghostly circles called palimpsests also tattoo Ganymede's surface. These are all that remain of impact craters that relaxed over time. Additionally, the moon has polar ice caps. To explain the thin, frosty coating of water ice over Ganymede's poles, we must consider what is happening inside the moon.

Unlike Callisto, which seems to lack a distinct rocky core, Ganymede's interior is highly separated. The giant moon has a hot, inner iron core, an outer core made of liquid iron, and a mantle of silicate rock. A thick layer of water ice surrounds all this rock and metal. It is the liquid iron inside Ganymede that generates a magnetic field, the only moon in the entire Solar System to do so.

The charged particles in Ganymede's magnetosphere interact with that of its parent planet, not only creating aurorae but bombarding its surface. These particles excite water ice molecules and cause them to deposit in the colder polar areas, creating the ice caps. If we could stand on the frosty polar vista, perhaps we could gaze at feathery aurorae dancing in the night, the land and sky both reminding us of what is churning deep under our feet.

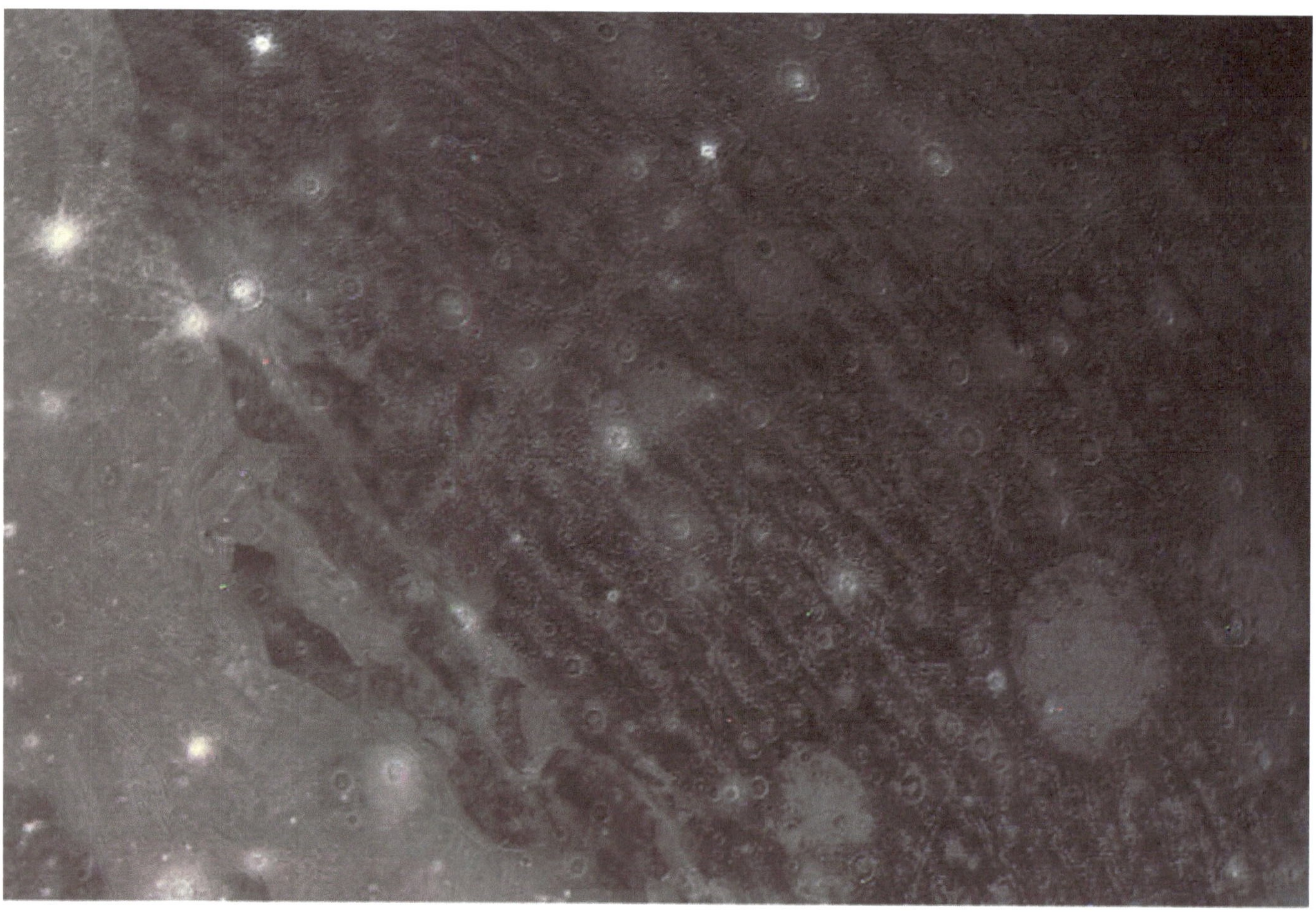

Furrows and palimpsests on Ganymede, as seen by NASA's *Voyager 2* spacecraft.

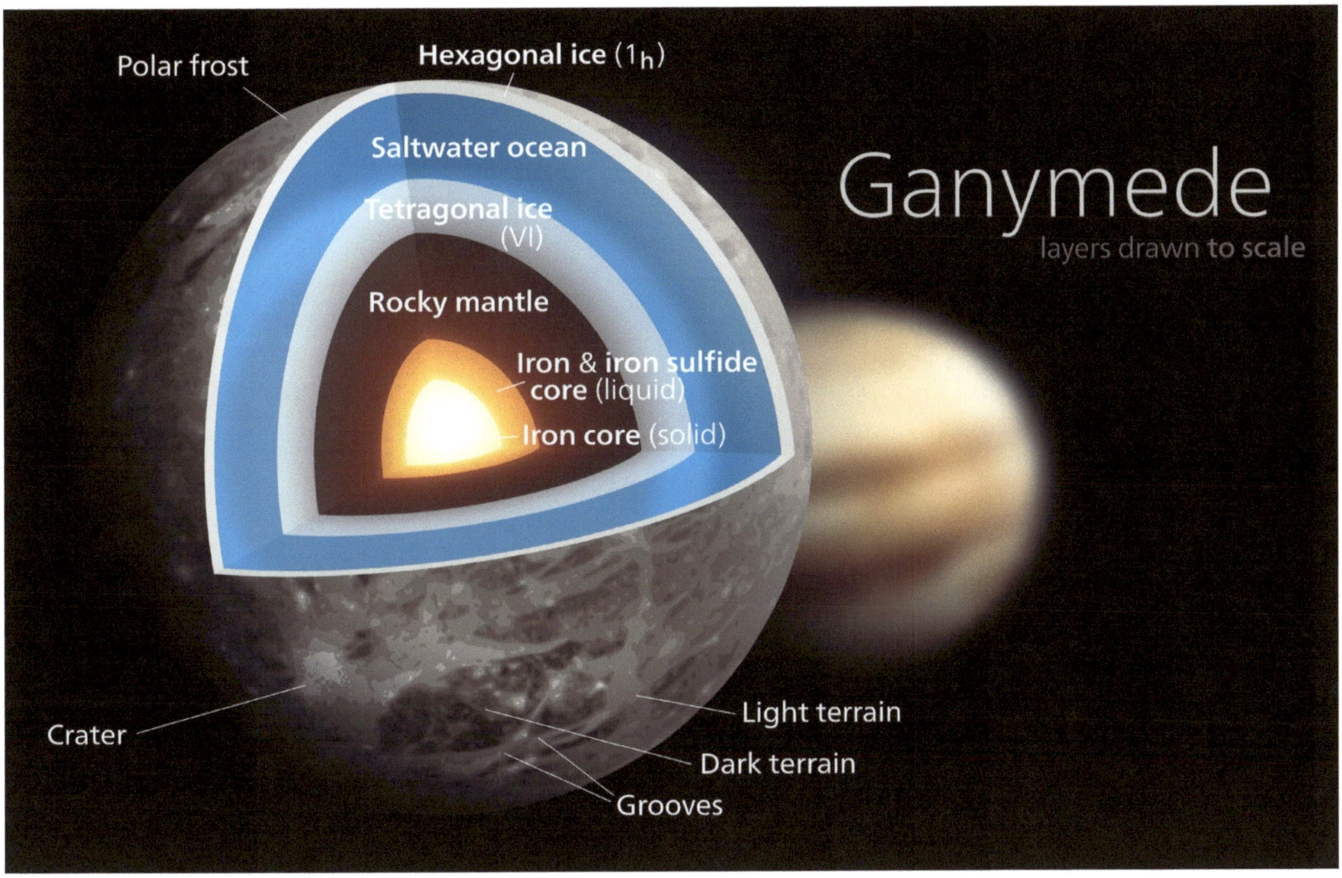

A visual model of Ganymede's internal structure.

The layer of water ice surrounding Ganymede's rocky interior is not all frozen. Gravitational measurements of the moon tell us that some hundred miles below the surface, the ice is melted into a layer of liquid water. This ocean itself may be hundreds of miles thick, containing more than 25 times the volume of all the oceans on Earth combined.

As the ocean is condensed deeper within Ganymede's interior, the liquid water turns into a high-pressure form of ice. Multiple layers of liquid water and high-pressure ice are perhaps stacked upon one another. Even if Ganymede's ocean is not directly in contact with the rocky mantle, plumes of hot material from its rocky floor could still seep through the high-pressure ice and enrich the ocean with minerals and nutrients. However, the extreme pressure of the ocean would suppress the chemical reactions needed for life to arise.

Poignantly, we may never know if the largest moon hosts alien microbes swimming in its waters. Since any life today would thrive hundreds of miles below a solid, rigid crust of ice, there is no obvious way to probe this mysterious underworld. As we continue our journey, we will find other oceans more accessible, and more promising to answer the question of whether or not we are alone in the universe.

Moon of Saturn

Iapetus

WHEN ASTRONOMER GIOVANNI CASSINI LOOKED at Saturn through his telescope in 1671, he discovered a new moon and named it Iapetus, after one of the titans in Greek mythology. But the moon seemed to appear and disappear as it orbited Saturn. Cassini correctly realized he had discovered a yin-yang world, dark as coal on one side and bright as snow on the other. Indeed, no other place in the Solar System offers us such a tale of two tones.

Iapetus is a strange work of art. Despite being a fairly large moon, just less than half the width of our Moon, Iapetus is made almost entirely of water ice and is a bit lumpy, with the poles visibly squashed. Its orbit is also peculiar. Not only does Iapetus orbit farther from its host planet than any other large moon, but unlike other regular moons, it circles Saturn at an incline of 17 degrees. The most likely explanation is that it was disturbed by a close encounter with another planet long ago.

Because of its inclined orbit, Iapetus is the only large moon from which we could clearly view Saturn's rings. In fact, all the moons closer to Saturn orbit in the same plane as the rings, and so from their surfaces the rings would always be edge-on and difficult to see.

NASA's *Cassini* orbiter captures Iapetus's two most unique features: its light and dark boundary and its equatorial ridge.

Iapetus's great distance from Saturn, however, puts it in the path of a huge, diffuse disk of dust. This dust originates from Phoebe, a dark and even more distant moon of Saturn. Phoebe, likely a captured asteroid or comet, has a relatively low mass and weak gravity. So, when meteoroids hit its surface, dusty material blows away into orbit around Saturn and is vacuumed by Iapetus, coating its leading hemisphere.

Indeed, Iapetus's leading hemisphere is the dark yin to the bright yang of the trailing hemisphere. But there is more to this story. If the two tones of Iapetus could be explained by a simple accumulation of debris from Phoebe, we would expect to see a gradual transition from the light to dark surfaces. Up close, this is not at all the case. In some places, it appears as if someone splattered a white canvas with splotches of black paint. In other areas, white spots overlaying dark land remind us of snow patches lingering in the shade, days after a snowfall.

What seems to happen is the blanket of dust heats up the leading hemisphere, causing ice to vaporize and deposit on the cooler trailing side. This darkens the leading hemisphere even more, so it gets warmer, and more ice vaporizes, in an endless loop that leaves no room for shades of gray. More than 300 years after Iapetus was discovered, we have finally cracked its black and white case.

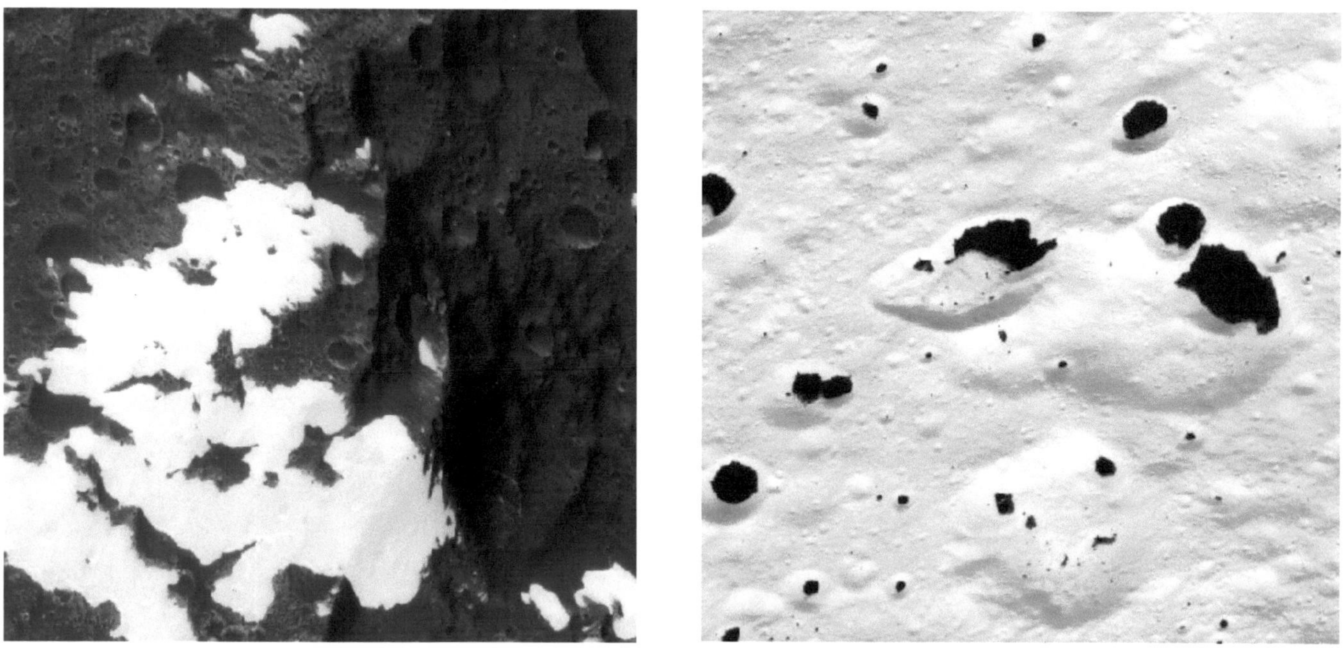

Two photograph from NASA"s *Cassini* orbiter showing (a) bright ice overlaying dark material, and (b) dark material splattered on the white surface.

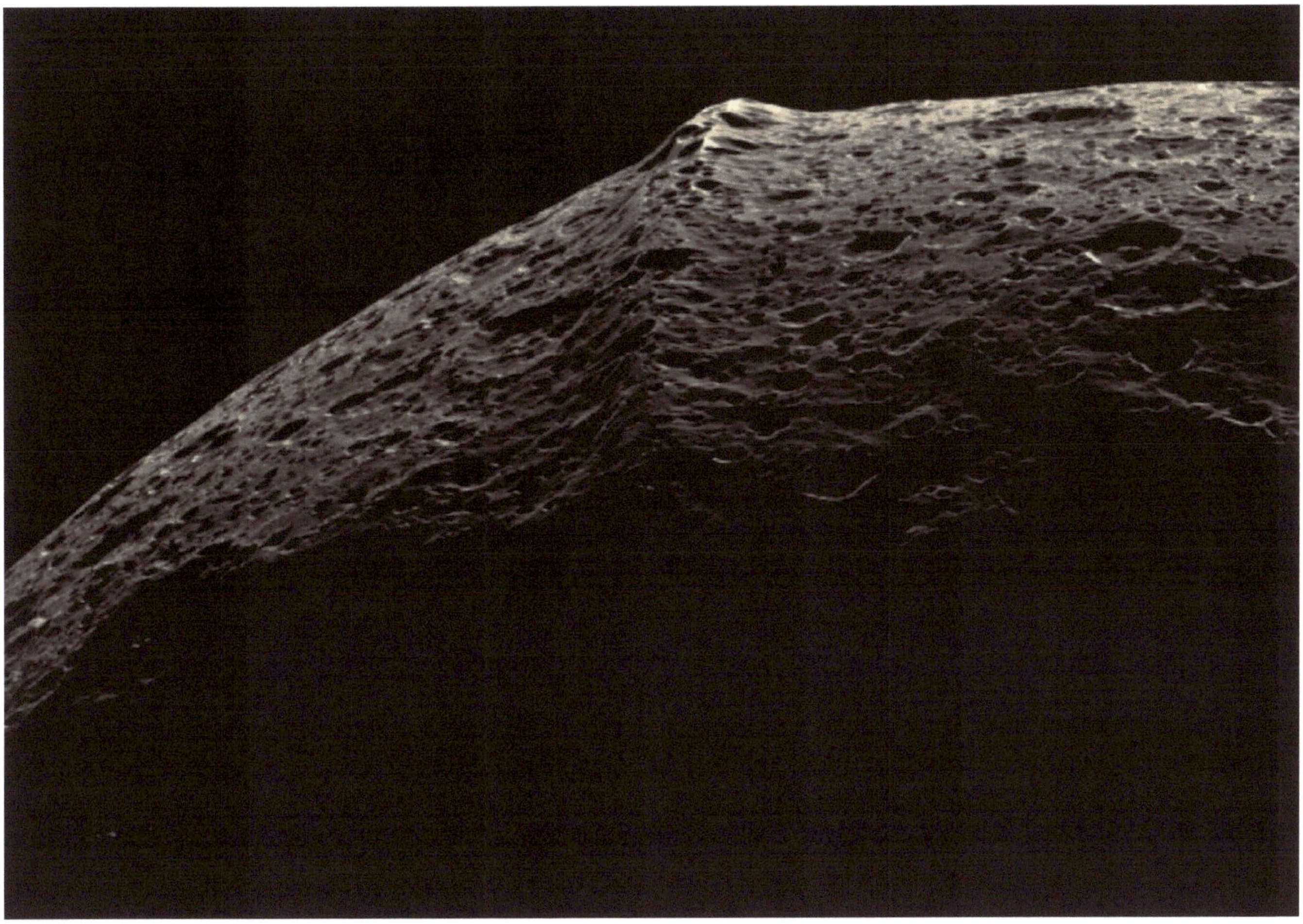

A crescent Iapetus with its equatorial ridge, as seen by NASA's *Cassini* orbiter.

Just as one mystery was solved, close-up photographs of Iapetus gifted us a new mystery: its equatorial ridge. Running along its equator for over three quarters of its circumference is a belt of mountains 800 miles long and so tall it gives Iapetus a distinctly walnut shape. Individual peaks tower up to 12 miles high, more than twice the height of Mount Everest. This spectacular landform appears ancient, but its origin is unknown. One interesting possibility is that it was created by the collapse of a former ring around Iapetus.

Regardless, the steep topography around these mountains, and surrounding Iapetus's many craters, lead to some of the most impressive avalanches in the Solar System, up to 50 miles long. Why they slide so far is not understood. Centuries after Iapetus confused its discoverer, the saga of trying to make sense of this moon continues.

Moon of Uranus

Miranda

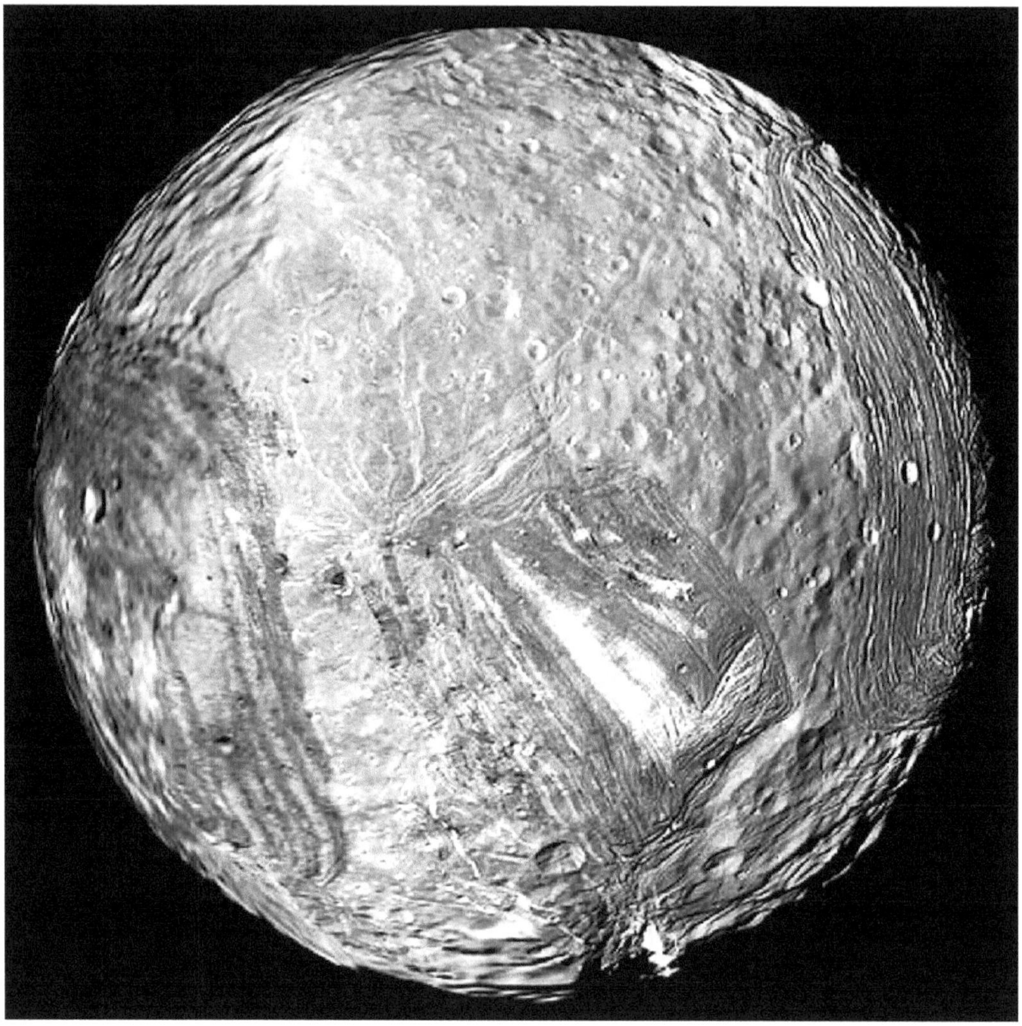

THERE ARE TIMES when nature presents us with something so unexpected, so outlandish, that we have to rethink everything and go back to the drawing board. Miranda, a moon of Uranus, is one such gift. Discovered in 1948 by astronomer Gerard Kuiper, and named after the character in Shakespeare's *The Tempest*, Miranda is a fairly small moon, about as wide as Iceland. So before Miranda was first imaged up close in 1986, it was assumed to be a boring, heavily cratered ball. How wrong we were.

Right: Elsinore Corona on Miranda, imaged by NASA's *Voyager 2* spacecraft.

Miranda is a world torn at the seams, with deep gashes and three racetrack-like, grooved regions called coronae. It looks as if someone shattered the moon and then randomly glued the pieces back together. The coronae of Miranda, each more than 100 miles across, stand out the most. The borders of one corona are particularly sharp, tracing a V-shaped pattern appropriately nicknamed the chevron. Another corona appears more freshly paved, reminiscent of the grooves on Ganymede, which hints its formation involved the flow of icy volcanic fluids.

Convection, the transfer of lighter material upward and denser material downward, possibly carried plumes of ice from Miranda's interior onto the surface, creating the unique coronae. How long this convection lasted, and why only Miranda bears witness to such a process, is a mystery.

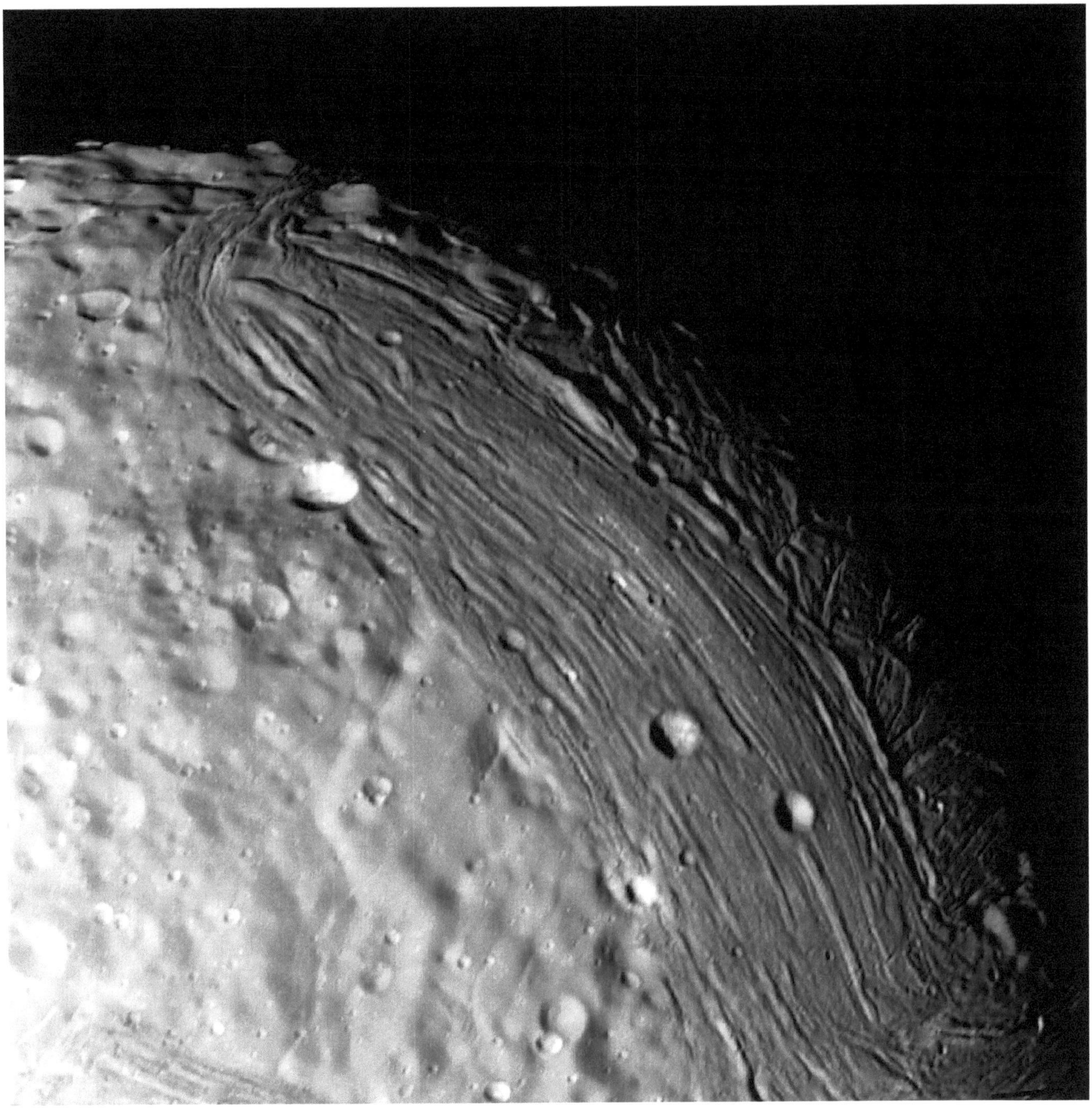

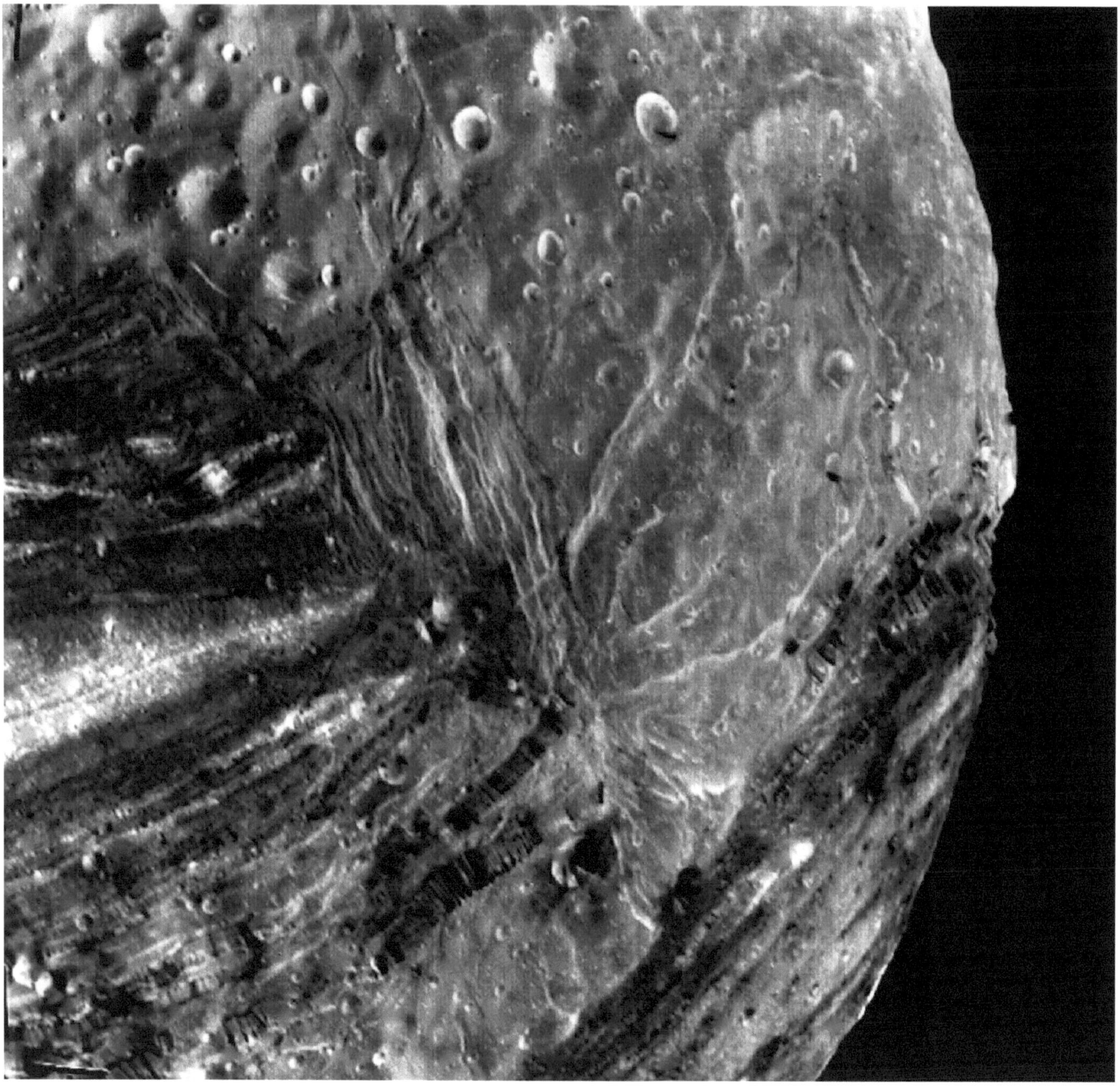

NASA's *Voyager 2* spacecraft captures faded craters between Elsinore and Inverness Coronae.

Surrounding the coronae are spectacular canyons and cliffs, tens to hundreds of miles long, slicing through the rugged landscape. The most prominent cliff, Verona Rupes, is up to six miles high, making it among the tallest cliffs in the Solar System. Because of its extreme height and also because of Miranda's low gravity, if we could jump off the edge of Verona Rupes, our descent to the bottom would take at least eight minutes. These canyons and cliffs were carved during a period of global expansion, perhaps around the same time the coronae formed.

The craters on Miranda are also strangely faded, indicating the moon is blanketed by a layer of dust, in places nearly a mile thick. This coating of dust, one of the thickest in the Solar System, may be ring debris that Miranda sweeps up as it orbits Uranus.

Miranda, a moon made mostly of water ice with perhaps a rocky core, is far too small to have powered all its geological activity on its own. Therefore, we must consider tidal heating. Miranda, some time ago, was likely in orbital rhythm with other larger moons of Uranus, in which every time Miranda orbited once, another moon orbited two or three times. This alignment kicked Miranda into a more elliptical orbit, so Uranus stretched it more when it was closer and less when it was farther. The repeated squeezing and releasing would have heated its interior, just as what happens when squeezing and releasing a stress ball.

Since Miranda is not currently in an orbital resonance, it is thought to be now frozen and dead. But extreme forces were clearly working on its surface relatively late in its history, perhaps one billion years ago. We marvel at how past tidal heating was able to deform such a tiny world. As our journey continues, we will be astonished to see what tidal heating, happening right now, can do on larger worlds.

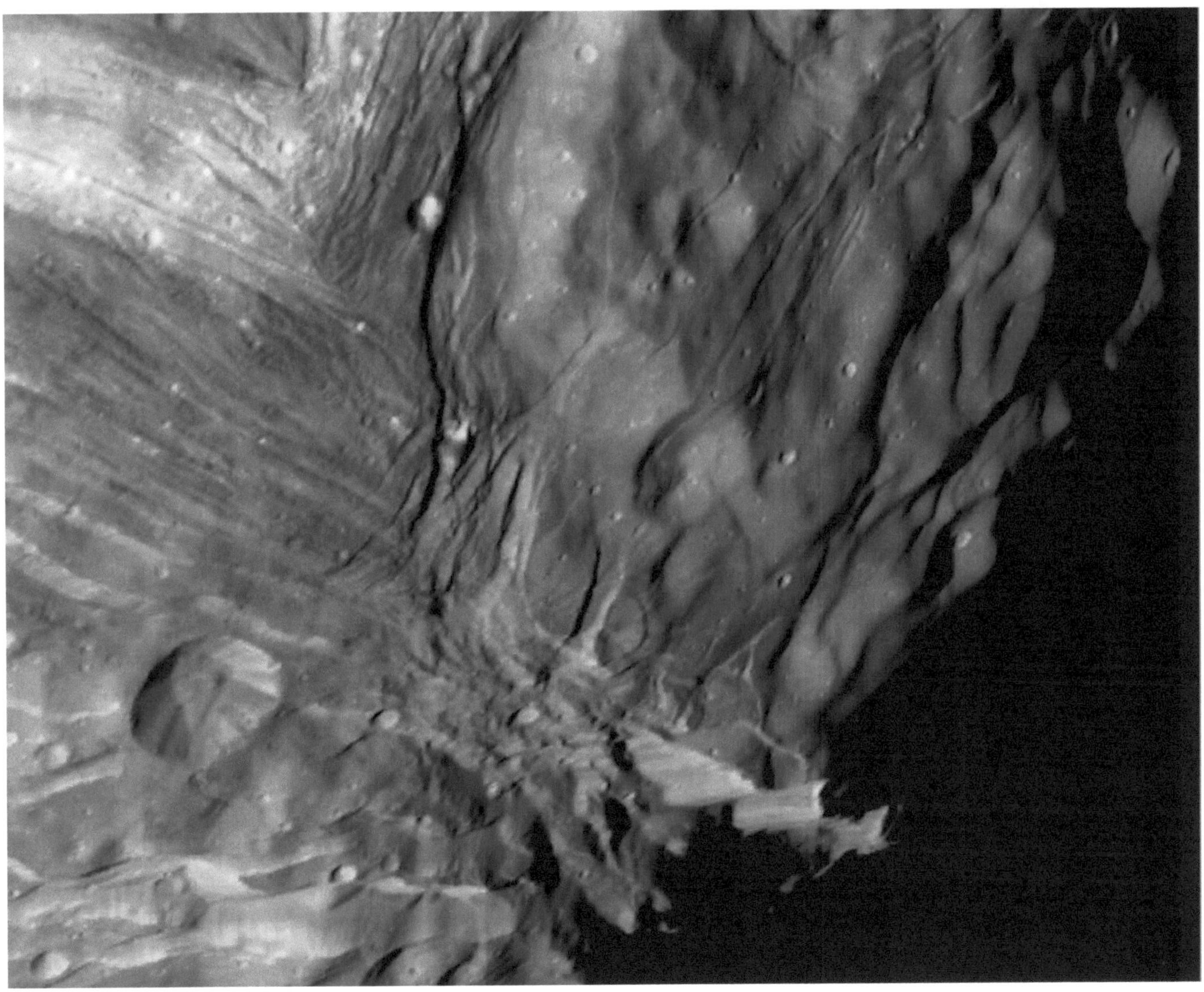

Verona Rupes at the bottom right, one of the tallest cliffs in the Solar System, as viewed by NASA's *Voyager 2* spacecraft.

Moon of Pluto

Charon

IN THE FRIGID REACHES of the Solar System, on an orbit extending well past Neptune, we find a romantic pair of icy spheres rotating around a common center of gravity. These worlds eternally keep the same face to one another, like two spinning figure skaters locked in arms. Later in this tour, we will explore the largest of the pair, Pluto. Our attention now turns to its smaller companion, Charon. At such a great distance from the Sun, Charon was expected to be a bland moon with little record of geological activity. Our first up-close encounter in 2015 proved otherwise.

About half the size of Pluto, Charon is one of the largest moons in the Solar System relative to the object it orbits. Discovered in 1978 by astronomer James Christy, Charon was the scientific-sounding name he suggested to honor his wife, Charlene. Ironically, in Greek mythology, Charon also happens to be the ferryman of the dead, closely associated with Pluto, god of the underworld.

Charon is a Frankenstein moon. On the side we have viewed up close, the northern hemisphere is rugged and ancient, whereas the southern hemisphere is younger and smooth. At the boundary of these two hemispheres is a belt of enormous canyons, stretching more than 1,100 miles across the surface. It looks as if someone cut two different moons in half and melded one half to the other.

The canyons on Charon are some of the deepest in the Solar System. One canyon, Argo Chasma, has walls nearly six miles deep, comparable to Verona Rupes on Miranda. For comparison, the Grand Canyon in the United States is around one mile deep. Rift valleys running along and near Charon's equator slice the crust into coherent plateaus.

Serenity Chasma cutting across Charon's equatorial region, imaged by NASA's *New Horizons* spacecraft.

Charon's plains, from NASA's *New Horizons* spacecraft. Kubrick Mons, a mountain in a moat, is to the right.

The southern hemisphere, at least on the side we have seen, is less cratered and flatter than the northern hemisphere. Most likely, this is a vast plain of solidified icy lava. Such lava flows were probably made of a slurry of water and ammonia. Isolated mountains, up to a few miles high and surrounded by moats up to a mile deep, pierce this landscape. These moated mountains do not appear to be volcanoes, but their origin is not certain, as they are unlike anything seen elsewhere in the Solar System.

Charon also has another distinctive feature: its red polar cap. While other worlds have polar ice caps, Charon's eerie reddish canopy is the only one darker than the rest of the surface. The reason for this bizarre polar cap involves Pluto itself. Because Pluto's gravity is fairly low, and because of how close Pluto and Charon are to each other, some methane molecules in Pluto's atmosphere escape and are captured by Charon. This methane then deposits as ice in the colder poles. Over time, solar radiation breaks down the accumulations into the frozen, organic gunk that coats the north pole, and presumably the unseen south pole as well.

Since Charon has a lot of impact craters, its tectonic and volcanic activity must have occurred several billion years ago. Perhaps heated by the impact with Pluto, Charon developed a rocky core and an internal liquid water ocean early in its history. With no other source of energy, this ocean froze, causing the moon to expand and rupturing its crust. The extensive network of canyons we see today are a relic from this violent past. The moon has since calmed down, but its vast plains, moated mountains, and deep gorges are forever etched on its face. And its poles, stained red by Pluto's atmosphere, remind us that the Earth and Moon are not the only worlds in the Solar System so intimately bound.

The pair of Pluto and Charon, as seen by NASA's *New Horizons* spacecraft.

Tormented by Tides

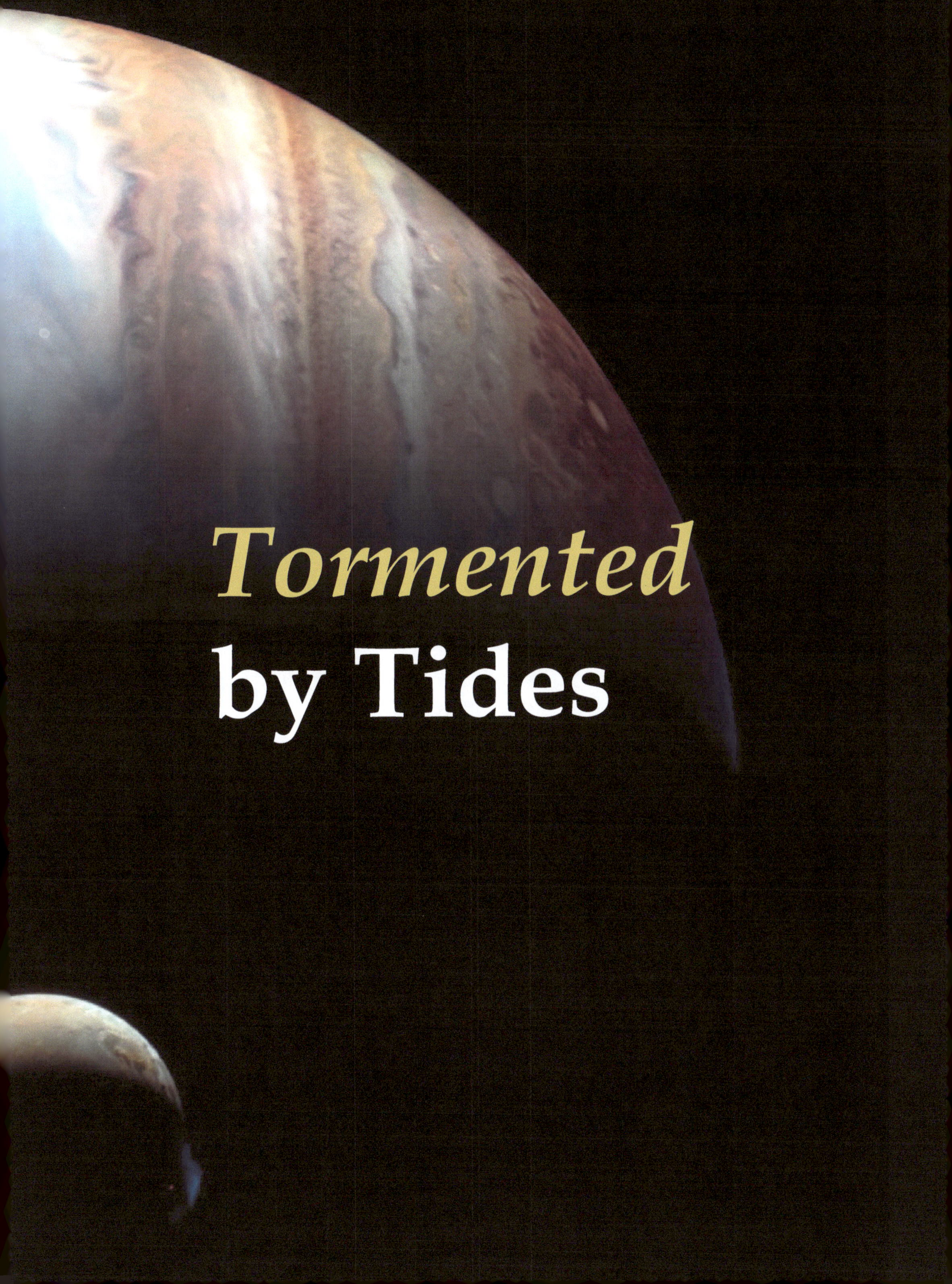

Moon of Jupiter

Io

WHEN JUPITER'S MOON IO was revealed up close in 1979, astronomer Linda Morabito at NASA wrote: *"I could feel tears begin to roll down my face at the sight of a world more unexpected than imagination itself. … It was as if I had been gazing into destiny."* She had just discovered the first volcanic eruption beyond Earth.

Right: NASA's *Galileo* orbiter captures volcanic plume of gas and particles jetting 60 miles into space.

Io is not like other worlds. Its face is painted with pastel shades of red, orange, yellow, and white, as if it were a cosmic pizza. This colorful moon, named for a female lover of Zeus in Greek mythology, has been at the center of two great scientific revolutions. First, Io was one of the four moons of Jupiter astronomer Galileo Galilei discovered in 1610, proving that not all heavenly objects revolved around Earth. Then, in the late 20th century, its fiery volcanoes dramatically heralded the power of tidal heating, and put to rest any notion that the moons of the outer planets would all be boring, frozen wastelands.

Similar in size and composition to our Moon, Io should no longer have enough warmth to drive geological activity. But in touring Miranda, we learned that tides can heat moons, and that is especially true of Io. This moon is locked in an orbital resonance with the neighboring moons Europa and Ganymede. Each time Io orbits Jupiter four times, Europa orbits twice, and Ganymede orbits once. These alignments tug on Io, keeping its orbit elliptical. So, as Io's orbit brings it closer to and farther from Jupiter, the gravitational pull of Jupiter stretches and squeezes the moon like a stress ball.

Such tidal forces on Io are astounding. On Earth, tides from the Moon lift the ocean by a couple feet. On Io, its entire crust rises and falls by 330 feet every 42 hours it completes an orbit around Jupiter. Tides heat Io's interior so much that a magma ocean may exist just 30 miles below the surface.

Loki Patera, the largest lava lake in the Solar System, as seen by NASA's *Voyager 1* spacecraft.

This tidal heating powers Io's volcanoes. More than 400 active volcanoes pepper its surface, with a variety of eruptions taking place at any given time. Not a single impact crater has been found, indicating its entire surface must be constantly replenished. Most Ionian volcanoes are not the large cone-shaped structures familiar to us on Earth, but are instead depressions called paterae. Some of the paterae are filled with lava, supplied from underground magma reservoirs.

The most colossal of these lava lakes is Loki Patera, 125 miles across and covering a greater surface area than Lake Ontario. No lava lake on Earth, or anywhere in the Solar System, is even close to this size. If we could stand atop the mile-high outer wall of Loki Patera and gaze to the horizon, we would behold an utterly alien scene: an endless sea of dark churning lava, perhaps with floating bergs of solid sulfur, and Jupiter ominously looming overhead.

Io also hosts the longest active lava flows in the Solar System. Rivers of lava, up to hundreds of miles long and at least three feet thick, stream across Io's surface. These originate when lava bubbles up from vents within paterae and then spills across the surrounding lands, a large-scale replica of how lava flows from the volcano Kīlauea in Hawaii. In 1997, lava flowed from a volcanic eruption into a nearby paterae, creating an astonishing "lava fall" two miles high and 40 miles long. Niagara Falls, less than 200 feet high and 800 feet wide, is a mere trickle in comparison.

Other paterae expel their lava with more fury. Enormous curtains and umbrellas of volcanic ash rocket up to 300 miles above the ground. Due to Io's lower mass and gravity, and also due to the sheer volume of the lava, Io's volcanic plumes reach far higher than those on Earth. If we happened to be on Io's surface at a safe distance, we would see fountains of bright liquids gushing up into the dark sky and blazing like flares. Since we would be in a vacuum, there would be no soundtrack to these dramatic shows, only haunting silence.

The sulfur and sulfur dioxide lava in these fountains immediately freezes and falls as snow, coating Io's surface in an assortment of warm-colored frosts. So, while temperatures in Io's flowing lava reach up to 3,000°F, hotter than any lava on Earth, the rest of its snow-covered surface is -230°F. Io is truly a land of fire and ice.

Tupan Patera, a lava lake on Io, imaged by NASA's *Galileo* orbiter.

Some of the mightiest mountains in the Solar System are found on Io. The tallest mountain, Boösaule Montes, is nearly 11 miles high, taller than Mount Everest. Unlike on Earth, where mountains tend to form chains, Io's peaks are isolated structures and often shaped like giant flat-topped tables, with steep edges. Because sulfur is not strong enough to shape such topography, these mountains must be composed of the bedrock silicates in Io's crust. The mountains of Io are not volcanic, but are formed by compressive stresses, which uplift chunks of Io's crust through tectonic thrust faulting.

Io is, literally, electrifying. As the moon orbits Jupiter, its escaped volcanic gasses interact with the giant planet's magnetosphere, becoming charged and tangled in the magnetic field lines. This sets up an electric current between Io and Jupiter, like an invisible cable more than 260,000 miles long. The current produces about two trillion watts of power, comparable to all the power humans produce on Earth. Jupiter's spectacular aurorae, and lightning within its atmosphere, are enhanced by this grand cosmic circuit.

Haemus Mons, rising six miles above the ground, captured by NASA's *Galileo* orbiter.

Picture of Io taken by NASA's *Juno* orbiter.

Io offers us some of the wildest wonders of the Solar System. And yet, our fascination with this moon will remain a long-distance affair for a while. Jupiter's magnetosphere bathes its surface in deadly radiation, five times the lethal human dose. That is to say nothing of all the violent volcanic eruptions, the extreme high and low temperatures, and the blanketing of toxic chemicals. One day, perhaps humanity will figure out a way to survive such a harsh environment, just to get a chance to bask in all of Io's glory. If so, Io will have yet again inspired a revolution in planetary science.

Moon of Jupiter

Europa

"All these worlds are yours, except Europa. Attempt no landing there."

THIS WARNING from Arthur C. Clarke's *2010: Odyssey Two* may be the work of science fiction, but we ignore it at our own peril. Europa, named for a female lover of Zeus in Greek mythology, and one of the four moons of Jupiter astronomer Galileo Galilei discovered in 1610, was part of a great revolution in astronomy. With the discovery of these four moons, no longer could anyone claim that every heavenly body revolved around Earth. But studying Europa up close for the first time in 1979 led to another scientific revolution. Just as Io dramatically proved that tidal heating can power volcanoes, Europa showed that tidal heating can also sustain oceans, and the possibility of life, well beyond Earth.

Europa is locked in the same orbital dance that keeps Io warm. Each time it revolves around Jupiter twice, Ganymede orbits once and Io orbits four times. These alignments pull on the moon, causing its orbit to stay elliptical. As Europa circles Jupiter, the giant planet's gravity squeezes the moon when it is closer, and stretches it when it is farther away. Since Europa is not as close to Jupiter as is Io, this tidal force is weaker. Still, its entire icy crust rises and falls by 100 feet with each orbit, far more than the ocean tides experienced on Earth.

Bearing witness to these tidal stresses, Europa looks like a bloodshot eyeball. Its smooth surface is bright and sparsely cratered, indicating it has been recently resurfaced, probably less than 100 million years ago. The vein-like features marring its face are long ridges and bands, up to thousands of miles long. The most common type of ridge on Europa is the double ridge, which consists of a pair of shallow cliffs separated by a sharp, thin trough in the middle. The exact cause of these ridges is unknown, but clearly involves tectonic fracturing of the ice shell. Other ridges on Europa are scalloped, as if someone scribbled a series of arcs on its surface. Such features likely formed when clean ice intruded into fractures, and their distinctive shape may correlate to tidal forces. Bands on Europa, which look like tire tracks in sand, are thought to form due to splitting of the brittle crust, with darker, warmer ice filling in the area between the brittle ice.

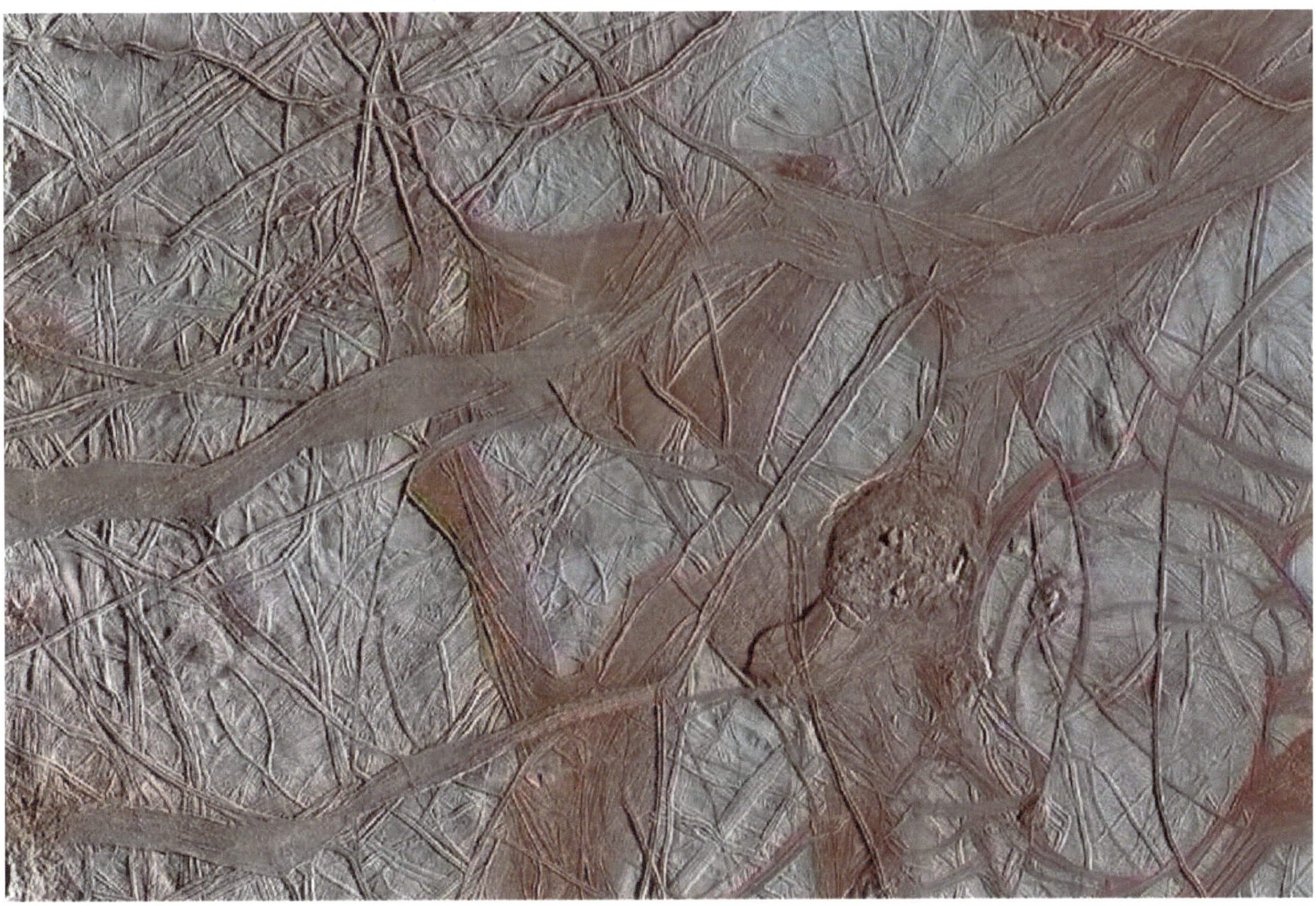

Red tectonic bands on Europa, as imaged by NASA's *Galileo* orbiter.

Illustration of Europa's subsurface ocea and likely plumes. Jupiter and Io are pictured in the background.

NASA's *Galileo* orbiter views Europa's heavily deformed terrain.

The heavily jumbled areas are aptly called chaos terrain. Some regions of chaos seem to be the tops of ice rafts which have twisted and turned around. Other areas of chaos are much smaller, appearing as reddish freckles dotting Europa's face. The origin of the chaos terrains is not certain. One possibility is that they are the result of warmer ice flowing upwards to the surface, similar to magma chambers on Earth, and then freezing in place like icebergs in a sea of solid ice. Another idea is that these disrupted regions sit atop lakes of liquid water, just a few miles underneath the solid ice. Either way, the chaos terrain of Europa strongly suggests that warm, flowing ice or even liquid water exists not far below the surface.

Europa's many tectonic markings, as well as gravitational measurements, indicate its ice shell is about 15 miles thick, and covers a 100-mile thick global ocean of salty, liquid water. The total volume of this ocean is likely two to three times greater than all the oceans of Earth, and itself overlays a rocky mantle and core of iron. The contact between the ocean and the rocky mantle is important for habitability, as rocky minerals sprinkled in the ocean can nourish hungry microbes. If hydrothermal vents exist on the rocky seafloor, they would be sources of energy and vital nutrients, as we observe deep in Earth's oceans. One difference is the extreme pressure at Europa's seafloor, five times greater than on the floor of Earth's oceans, which may make it more difficult for life to sustain itself around such vents.

Space telescopes have discovered what appear to be plumes of water vapor shooting 200 miles above Europa's surface. While this may stir dreams of sending spacecraft to sample the material and look for signs for life, they seem to only occur periodically, which will complicate future sampling efforts. Moreover, the plumes may not originate directly from its mineral-infused ocean. Given the 15 miles of solid ice and the relatively high gravity of Europa, it would take an extremely high amount of force to power such eruptions. Rather, the source could be shallower lakes, which are encased within the ice crust and not thought to be connected to the deeper ocean below.

Europa is a world with many promising signs in our search for life beyond Earth. Heated by tides, it has a global water ocean under a relatively thin layer of ice and in contact with a warm, rocky mantle. And yet, Europa still presents challenges. Like Io, Europa is bathed in Jupiter's deadly radiation, which would kill any humans walking on its surface within hours. In fact, studies hint that the radiation interacts with the salts on its surface and causes the moon to glow in the dark, even when faced away from the Sun. Without radiation protections, we would be wise to heed Arthur C. Clarke's warning and attempt no landings on such a dangerous place. Fortunately, our next destination is more inviting.

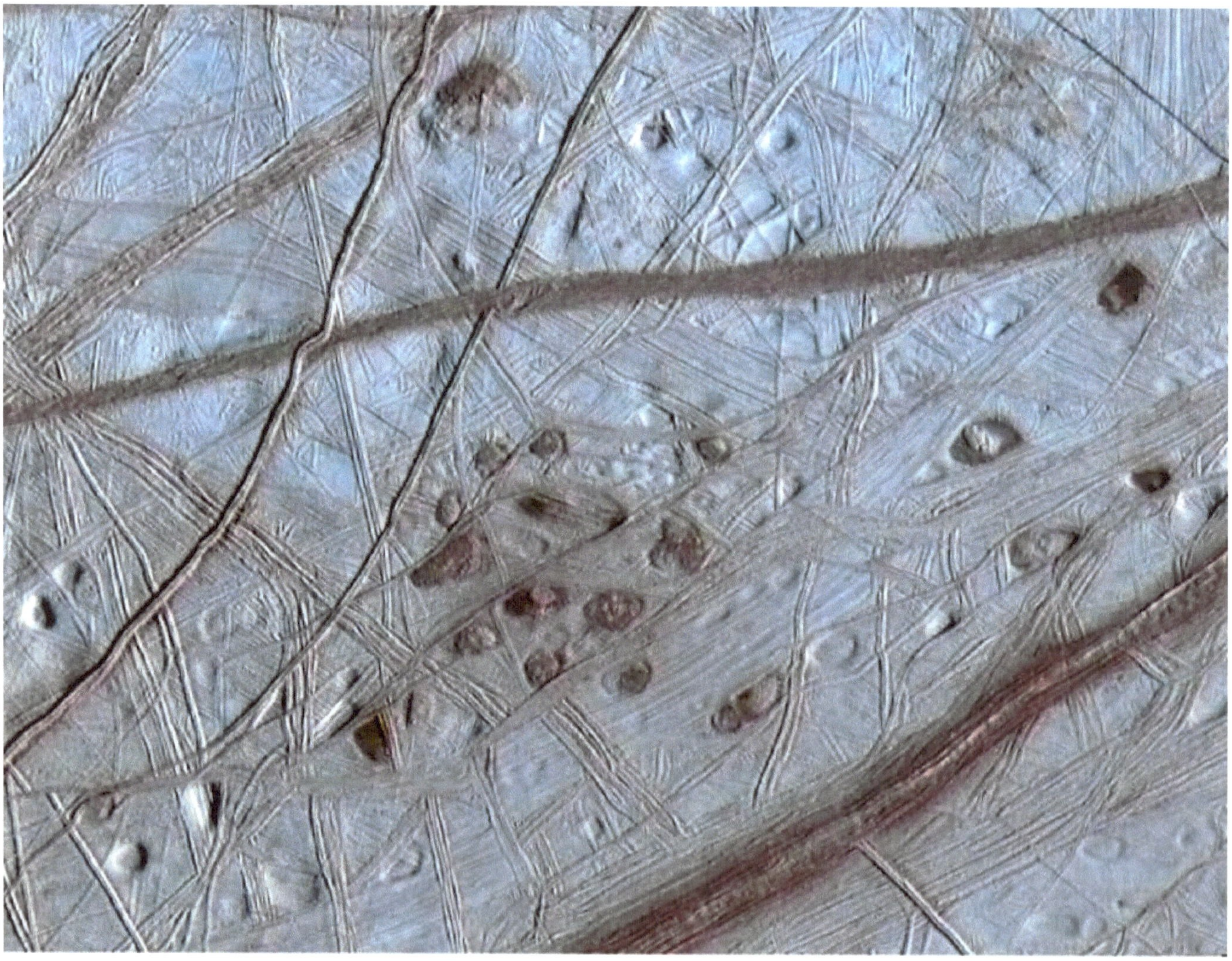

Spots dotting Europa's icy surface, seen by NASA's *Galileo* orbiter.

Enhanced color view of Europa and its many tectonic markings by NASA's *Galileo* orbiter.

Moon of Saturn

Enceladus

IT IS IRONIC that Enceladus, discovered in 1789 by astronomer William Herschel, was named after a giant in Greek mythology. Enceladus is a rather small moon of Saturn, comparable in size to the United Kingdom. But Enceladus towers in other ways. If we are to ever find alien life which arose independently from life on Earth, Enceladus is our best place to look in the Solar System.

Right: Grooves and fractures on Enceladus's surface, as photographed below by NASA's *Cassini* orbiter.

Up close, Enceladus is paved with wide grooves, similar to those seen on Ganymede. But unlike on Ganymede, all of Enceladus is the same white color. The moon also features canyon networks, with lengths reaching more than 100 miles and depths of close to a mile. The grooves and canyons of Enceladus are considered to be tectonic faults formed from the stretching of its crust. Water or slushy ice erupted from the interior to fill in the widest fractures, creating the grooves we see today. Based on the lack of many craters, these features are very young, less than 100 million years in age.

Enceladus is the brightest world in the Solar System, reflecting nearly all the light it receives from the Sun like a cosmic mirror. In fact, the entire moon is covered in freshly fallen snow. But the snow does not come from clouds, as Enceladus has no atmosphere. The snow comes from inside Enceladus. In the south pole, more than 100 jets of water erupt from parallel fractures called the tiger stripes. About 550 pounds of water per second gushes out of these jets at up to 1,360 mph. Due to Enceladus's low mass and weak gravity, the jets reach more than 6,000 miles into space.

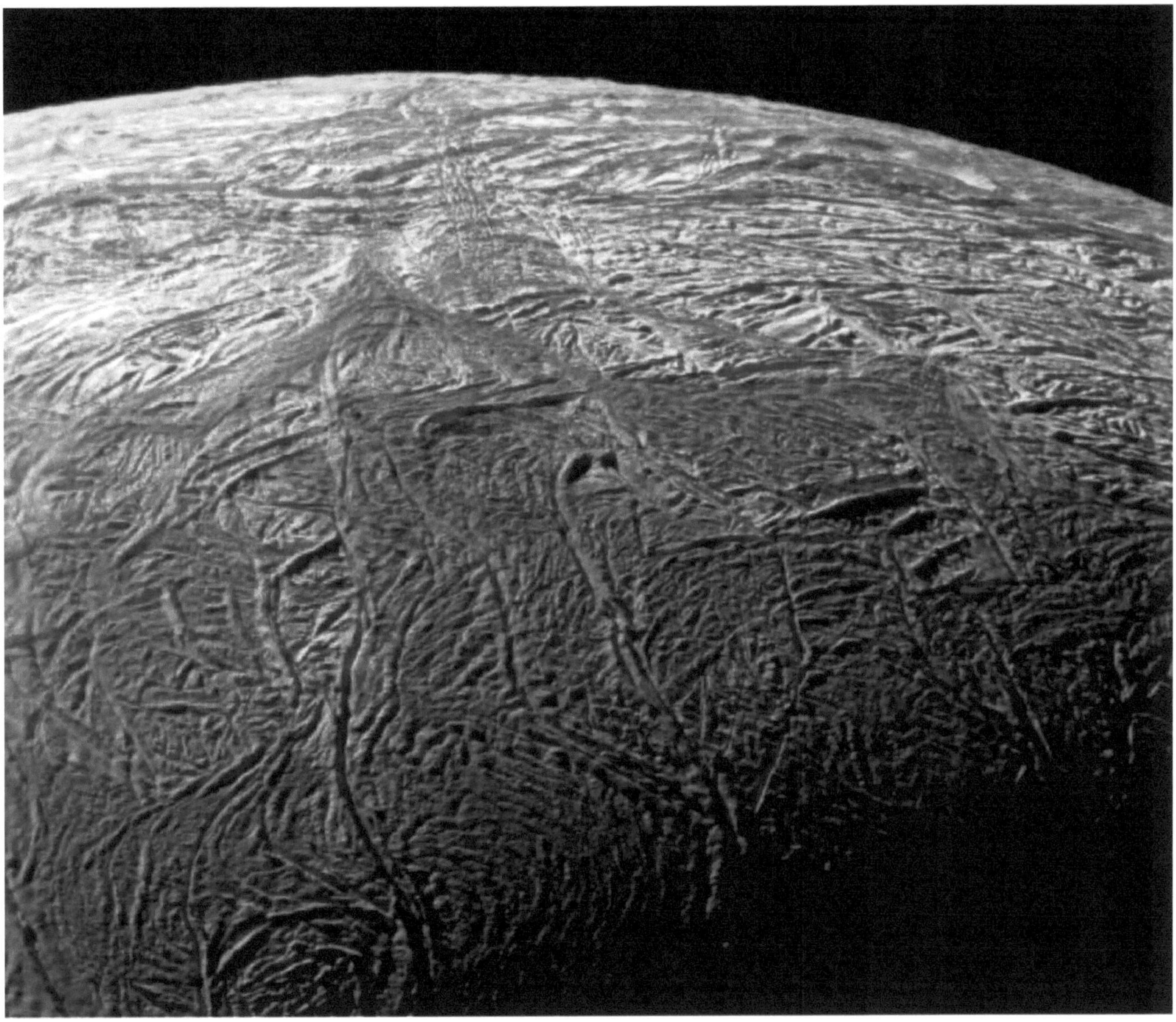

The water particles in the jets immediately freeze, some of which fall back onto Enceladus and coat its surface in fine, powdery snow that would be a skier's paradise. Snow piles on Enceladus can reach up to 2,300 feet deep. The rest of the plume material escapes into space and forms one of Saturn's rings, the wide and ethereal E ring. Microscopic snow droplets within this ring reach and brighten several other nearby moons of Saturn, and even fall into Saturn itself. Enceladus is truly the Solar System's snow cannon.

Pictured: The water fountains of Enceladus, as seen by NASA's *Cassini* orbiter.

The fountains of Enceladus also point to subsurface liquid water. Gravity measurements of the moon indicate it has a global water ocean just 5 to 15 miles underneath its icy crust. This ocean, estimated to be about 25 miles thick, envelopes a rocky core. Because of Enceladus's low gravity, its plumes are able to eject water directly from this ocean, not from subsurface lakes as is thought to be the case on Europa. The plumes on Enceladus are also constantly erupting, which will make future sampling missions quite simple.

How tiny Enceladus has the power to maintain this ocean and activity is baffling. Like Io and Europa, Enceladus is in an orbital resonance with a neighbor moon. Every time Enceladus orbits Saturn twice, the moon Dione orbits once. This tugs on Enceladus and keeps its orbit elliptical, which warms its interior as it orbits and feels the squeezing and stretching of Saturn's gravity.

But Enceladus is much smaller than Io and Europa, and Saturn's pull is also not as strong as Jupiter's, so it should not be as active. One compelling explanation is that unlike those worlds, Enceladus's core is porous, more like a sponge. Such a soggy core can be stretched and squeezed far more easily than a nonporous core, which multiplies the amount of tidal heating it receives. A porous core also allows water to seep into it, transferring heat and chemicals.

In fact, hydrogen was detected in the plumes of Enceladus, which almost certainly formed through hydrothermal vent activity at the ocean floor. Carbon, nitrogen, phosphorus, oxygen, and organic compounds have all been detected in the plumes, proving that the ocean is enriched with vital nutrients. With water, energy, and the right chemistry, Enceladus seemingly has all the ingredients needed for microbial life.

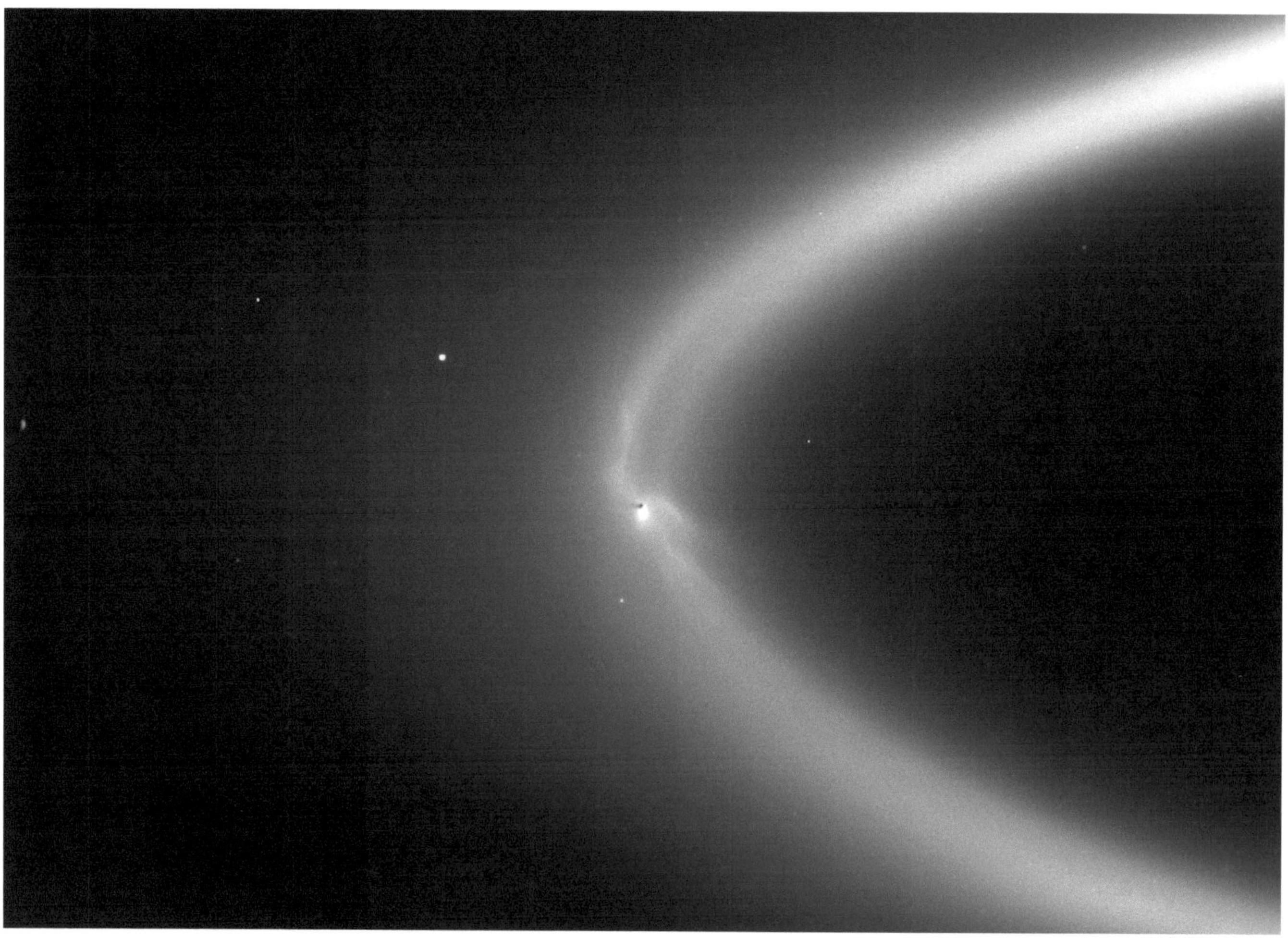

Enceladus is embedded in Saturn's E-Ring, which it supplies with its water plumes, as seen in the below image from NASA's *Cassini* orbiter.

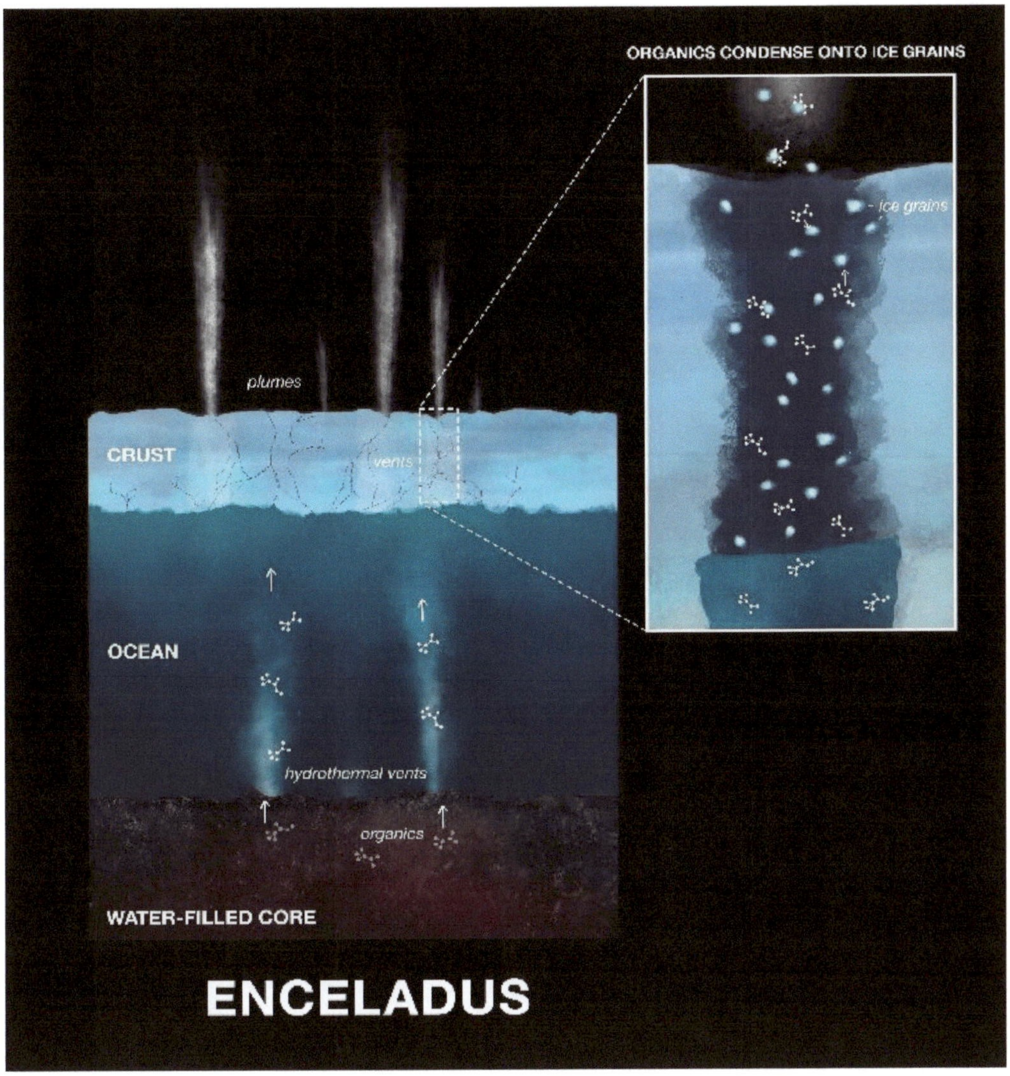

Illustration of how organics are transported into Enceladus's ocean and plumes from hydrothermal vents.

The major question, however, is how long the ocean has been maintained. Saturn's rings must have formed from a cataclysmic event, but when this occurred is uncertain. Therefore, the age of Enceladus's ocean and even Enceladus itself is also not clear. Because of its small size, we cannot assume the ocean is as old as the moon. Current theories suggest the ocean is about one billion years old, which is long enough for life to arise, but more studies are needed.

Enceladus, the glistening gem of the Solar System, invites us to look closer. Unlike Europa, Enceladus is not tucked within a deadly radiation environment. The constantly erupting plumes, the clear evidence for hydrothermal activity, and the detection of complex organic chemistry make this moon a top target for future exploration. Someday, even humans could walk its lands. We can imagine making footprints in the snow, admiring from a safe distance the epic fountains and Saturn hanging in the sky like a gold ornament.

But most excitingly, this little world near Saturn's rings may hold the answer to our greatest question: are we alone?

Under Alien Skies

Terrestrial Planet

Venus

LOOKING TO THE EAST near sunrise, or to the west near sunset, you may notice a bright, pearly jewel sparkling in the sky. This is Venus, the closest planet to Earth and the second planet from the Sun. Named after the Roman goddess of love and beauty, Venus is nearly the same size, mass, and composition as Earth. Before it was explored up close, Venus was imagined to be a twin of Earth, a warmer, tropical paradise. We now know the truth: something went terribly wrong.

The first clue pointing to Venus's different evolution is its atmosphere, which is mostly composed of carbon dioxide. The atmospheric pressure is more than 90 times that of Earth's, similar to the pressure we would experience 3,000 feet under Earth's oceans. This noxious greenhouse blanket traps solar radiation and warms the planet's surface to 860°F, hot enough to melt lead, making Venus the hottest planet in the Solar System. The thick atmosphere is such a drag on Venus's rotation that the planet actually rotates backwards. One day on Venus is equal to 243 days on Earth. At the end of the day, if we were unfortunate enough to be on the planet, we might see through the clouds the dimly glowing Sun setting in the east.

Just as Earth's atmosphere is mostly composed of nitrogen but has clouds of water vapor, Venus's carbon dioxide atmosphere has clouds of sulfuric acid. These clouds reflect around 70% of the sunlight that reaches them, and since they cover the entire planet, it is impossible to see Venus's surface from space. Sulfuric acid rains from these clouds, but because Venus is so hot, the rain evaporates 10 miles above the ground. When a planet is too extreme for even sulfuric acid to survive, we know humans would stand little chance.

NASA's *Magellan* orbiter peered through the clouds of Venus and imaged its surface using radar, as pictured above.

And yet, Venus has a remarkable surface. Radar imagery has allowed us to see through the clouds and find a world saturated with at least 85,000 volcanoes. The volcanoes of Venus come in many different shapes and sizes. Several are shield volcanoes, similar to the volcanoes of the Hawaii islands. The largest is Maat Mons, rising three miles above the ground and stretching 245 miles across. Other volcanoes are unlike any seen on Earth, such as the ones resembling giant pancakes up to 60 miles wide. The origin of these bizarre structures is not quite understood, although probably involves extrusion of magma from underground chambers.

Solidified lava rivers also snake around Venus's surface. Baltis Vallis, a narrow river of solidified lava, is 4,200 miles long, the longest channel in the Solar System. If a lava channel of that length flowed from the Kīlauea volcano in Hawaii, it would reach the Mississippi River. Another type of unusual feature found on Venus is its coronae. Typically hundreds of miles across, the coronae of Venus appear as circular structures surrounded by a crown of fractures. These coronae likely developed due to a combination of tectonic faulting and volcanic eruptions along the crowns.

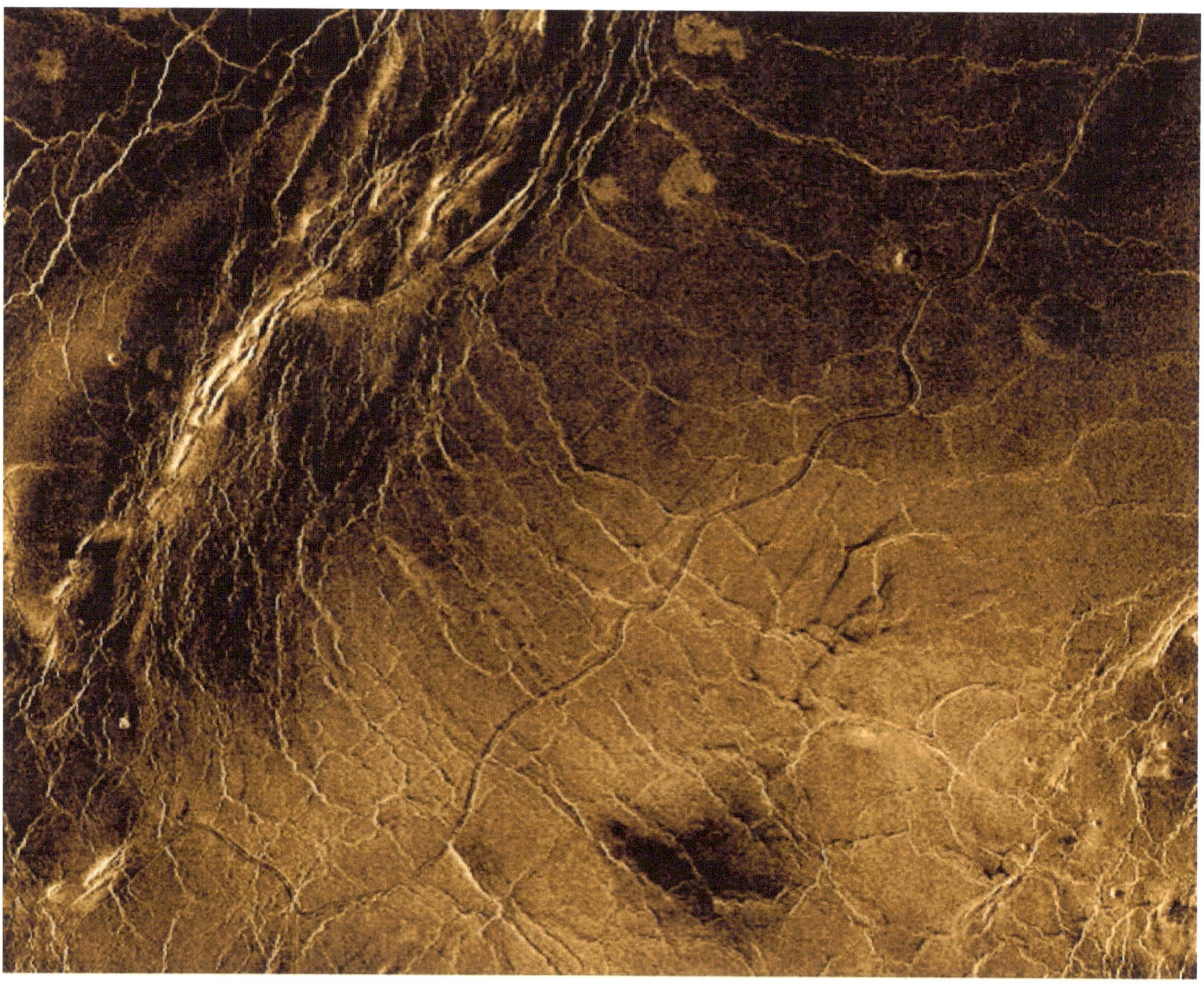

Radar image of Baltis Vallis, from NASA's *Magellan* orbiter.

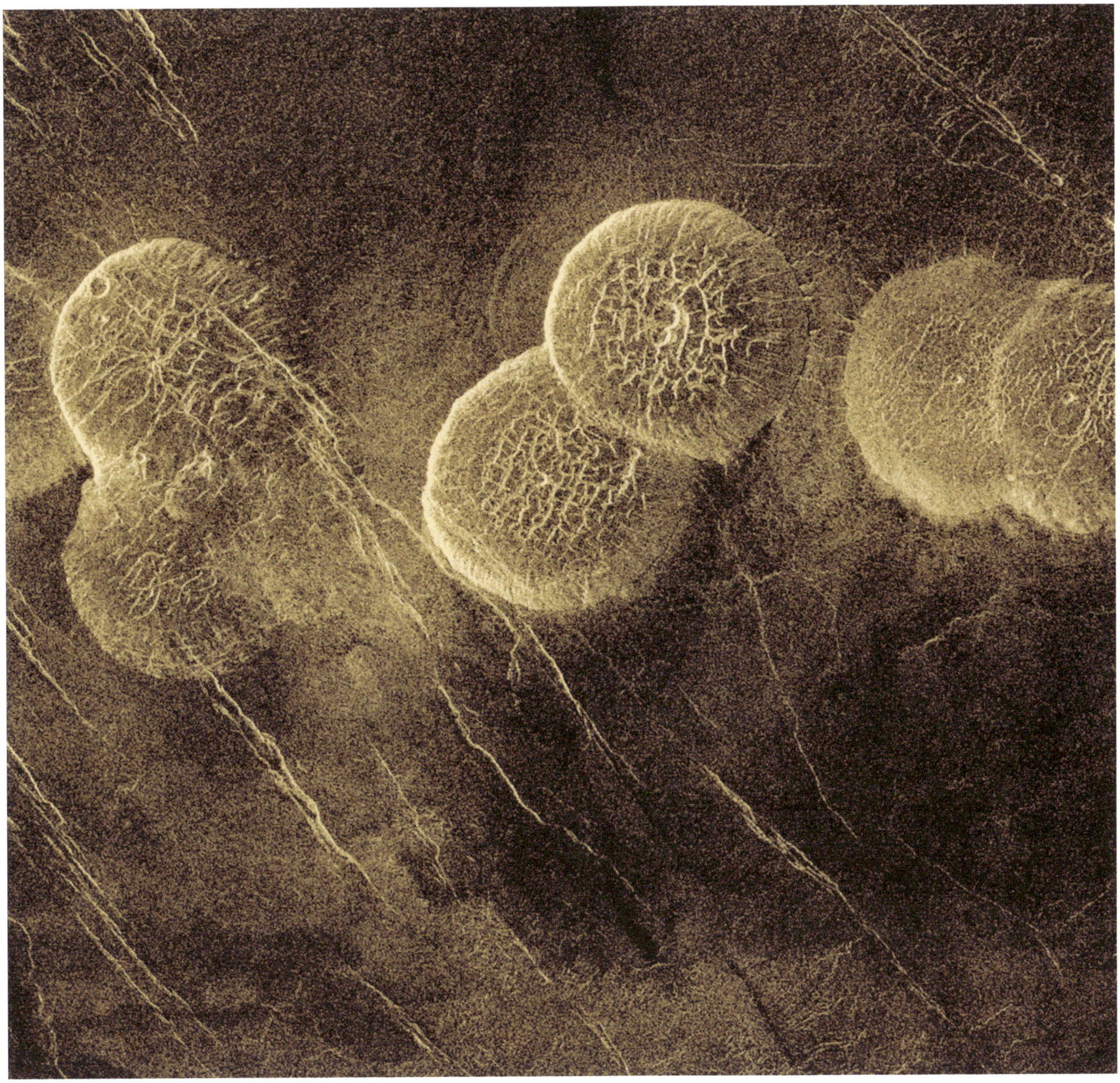

Radar image from NASA's *Magellan* orbiter of pancake domes, weird volcanoes found only on Venus.

Tectonics have clearly played a large role in shaping Venus's surface. Rift valleys up to 2,500 miles long and three miles deep cut across Venus, the result of crustal faulting in connection to the formation of massive shield volcanoes. Tectonic processes have also created mountain chains on Venus. The tallest mountain, Maxwell Montes, towers four miles above the surrounding plains. The top of Maxwell Montes and other mountains are capped in bright material, which may be deposits of metal "snow."

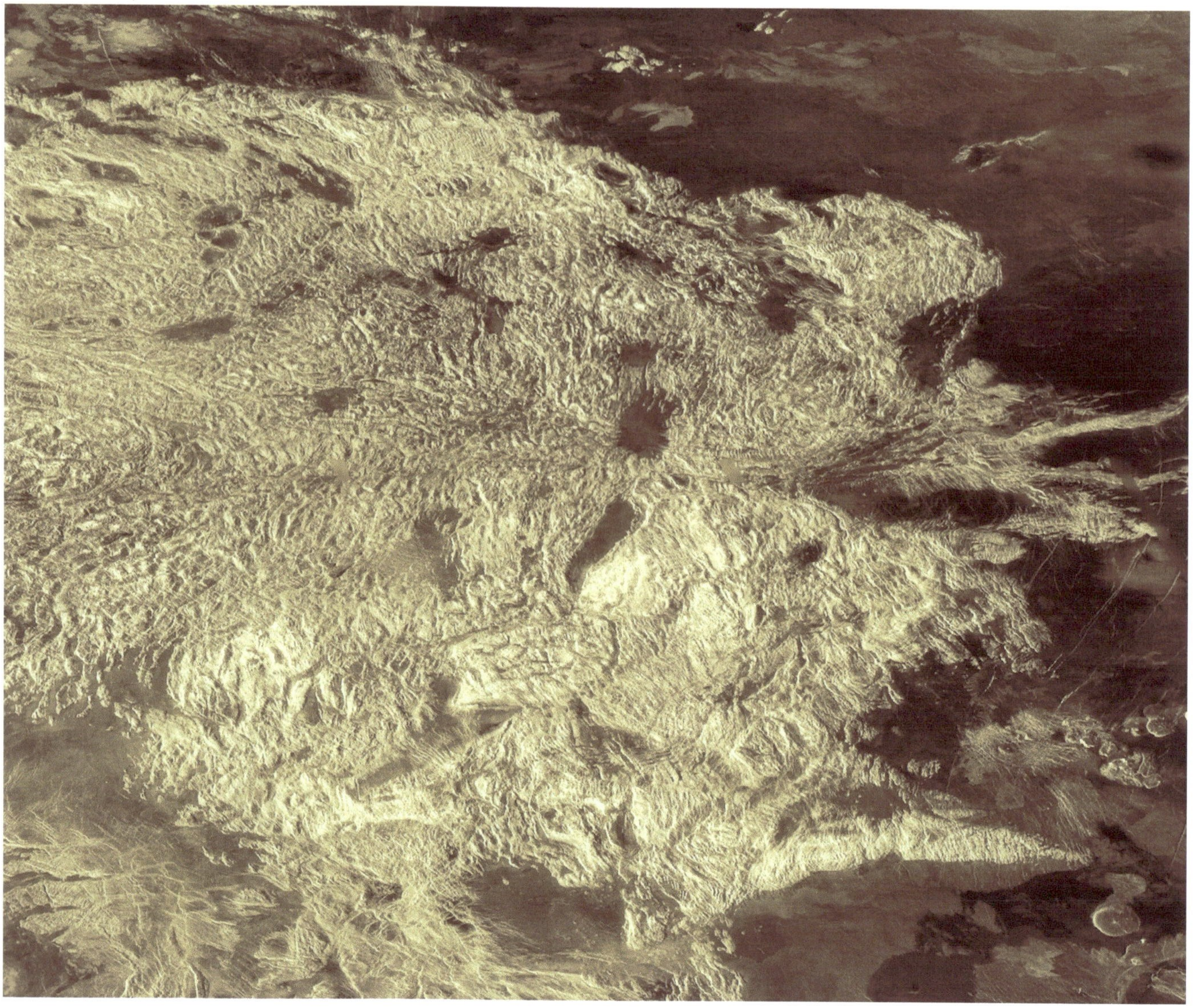

Bright tesserae on Venus, as viewed in radar imagery from NASA's *Magellan* orbiter, may be the oldest features on the planet.

The key to this history may be in the heavily deformed, folded hills called tesserae. The origin of the tesserae is uncertain, but they appear to be the oldest terrain on the planet, which makes them potential time capsules. Perhaps within these tesserae are old pieces of crust which were modified by past liquid water.

Because Venus's entire surface is sparsely cratered, most of its volcanic and tectonic activity probably occurred in the last 500 million years. Thermal observations strongly suggest lava is still erupting and flowing on Venus today. The universally young surface complicates our efforts to understand how Venus evolved to become so different from Earth. Venus may well have started out a wet planet like Earth, with lakes and oceans, but when this water evaporated, or whether it existed at all, is a mystery.

Could alien lifeforms still exist in Venus's clouds? The question might seem preposterous, but approximately 30 to 40 miles above the ground, the temperature and pressure are similar to that of the surface of Earth. Studies hint that microorganisms, lacking any water, could thrive off the sulfuric acid droplets within the planet's upper atmosphere. Such a possibility, while remote, forces us to consider that exotic, waterless environments may not always be as inhospitable as we once assumed.

In any case, the surface of Venus is a foreboding place, a cautionary tale for our own future. If Venus was once the tropical paradise of our earlier imaginations, its scorched ground is now a vision of hell. Finding out why Earth's twin suffered a runaway greenhouse effect is one of the most important questions we can answer as we deal with the ramifications of our own greenhouse experiment. Due to the extreme surface conditions and the rampant volcanism, Venus does not make our task easy. The next planet on our tour may have also had oceans, but has preserved much more for us to explore.

Venus in two different lights: infrared on the left, and ultraviolet on the right, as processed from Japan Aerospace Exploration Agency's *Akatsuki* orbiter.

Terrestrial Planet

Mars

MARS IS THE MOST OMINOUS of all the planets visible to the naked eye. Named for the Roman god of war, Mars looms like a warning sign, an unblinking red light in our night sky. When telescopes were invented, astronomers pointed them at Mars and found polar ice caps and what were interpreted as long canals. Some thought they may be the engineering work of a civilization drying up, looking jealously upon Earth and trying to transfer water from the poles to save themselves. In the 1970s, finally seeing Mars up close, we realized the planet is rusted and uninhabited. The only civilization in the Solar System is ours.

But Mars has many wonders. About halfway in size between the Earth and Moon, the red planet is just massive enough to hold onto an atmosphere, mostly of carbon dioxide. Due to differences in light scattering, sunrises and sunsets on Mars are blue, while the daytime sky is a dusty tan color. The atmosphere of Mars is thin, less than 1% the pressure of Earth's atmosphere, but it is capable of whipping up giant dust storms with winds in excess of 60 mph. As sunlight heats different parts of the surface, the energy imbalance kicks up dust and initiates the storms.

On Earth, dust storms are limited by water, and water-supported regions of vegetation. Because no bodies of liquid water exist on the surface on Mars, its dust storms can grow unimpeded, sometimes covering the entire planet. Such planet-encircling storms occur about once every three Martian years. If we happened to be on Mars and caught in a dust storm, we would probably feel nothing more than a gentle breeze, given the low air pressure. However, the dust particles would cling onto our space suits like styrofoam packing peanuts.

Martian dust storm, from NASA's *Mars Global Surveyor* orbiter.

Mars is the only world in the Solar System, besides Earth, where we know snow falls from clouds. However, snow on Mars is different. While the daytime summer temperatures near the equator can reach up to 70°F, nighttime winter temperatures at the poles dip to -190°F. In these chilly nights, microscopic snowflakes of carbon dioxide, also known as dry ice, occasionally reach and even coat the ground. Standing in the middle of this snow with night-vision goggles, we would appear to be enveloped in fog, and it would be utterly quiet. At dawn, perhaps we could take a brisk walk in fields of this fine, powdery snow.

The polar ice caps of the Martian poles are permanent features. Both caps are about one mile thick, made of water ice, and blanketed by frozen carbon dioxide. The deposits of carbon dioxide vary with topography, and even slowly flow downslope as glaciers. The thin layer of carbon dioxide forms with the seasons. Similar to Earth, Mars is tipped on its axis by 25 degrees, so as one pole moves to winter, nearly a quarter of the carbon dioxide gas in the atmosphere freezes and coats the polar water ice. During spring, sunlight heats the layers below the top of the ice. The resulting buildup of pressure causes geysers which shoot carbon dioxide gas 250 feet in the air and last no more than two hours. These solar-powered springtime geysers are spaced roughly a mile apart and cluster across hundreds of miles.

The flow of a past glacier, pictured by NASA's *Mars Reconnaissance Orbiter.*

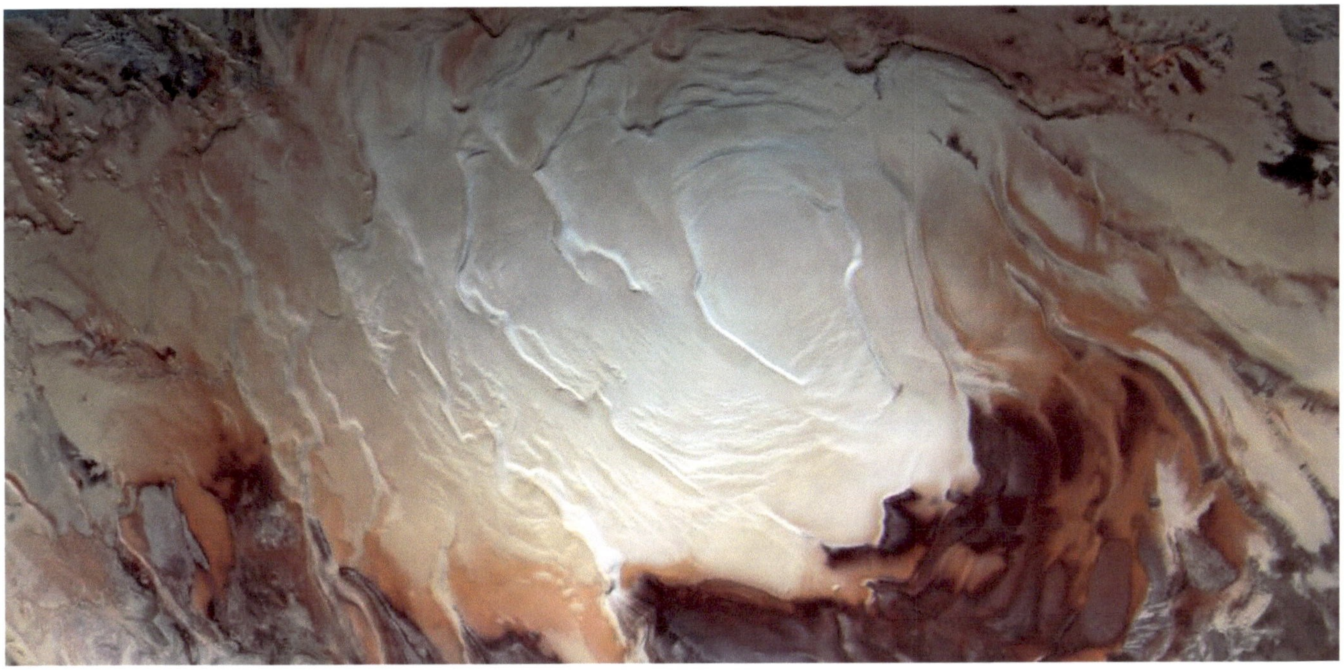

The southern polar cap of Mars, as seen by ESA's *Mars Express* orbiter.

Korolev Crater, a natural ice-skating rink in Mars, photographed by ESA's *Mars Express* orbiter.

Dark sand dunes on Mars, as seen by NASA's *Curiosity* rover.

Like Earth, Mars experiences ice ages as its axis wobbles. In its mid-latitudes, moraines and lobate flows are recognized as the remnants of glaciers from roughly two million years ago, when the Martian climate was much colder. Belts of glacial deposits, buried under at least a few feet of rocky debris, have also been located across these areas. Another marker of the planet's climate changes are its dunes. Covering a small percentage of the surface, the dunes of Mars are nonetheless strangely beautiful. Some dunes are circular, like fuzzy plush balls sitting atop a sandbox. Other dunes are more familiar. The largest sand sea on Mars, similar in size to the sand sea of the Arabian Peninsula, circles its northern ice cap. Dunes here reach up to 80 feet high, and other isolated dune fields on the planet can reach up to 1,000 feet tall.

Mars boasts the largest volcano in the Solar System. This volcano, named Olympus Mons, is 375 miles across, nearly as wide as France. It is also 13 miles high, much taller than any mountain on Earth. However, due to its large width, the slopes are gradual. In fact, if we stood on Olympus Mons, we probably would not be able to tell we were on a volcano.

Olympus Mons is part of a colony of massive shield volcanoes, all of which seem quiet now. However, these are not ancient features. The last eruptions on Olympus Mons likely occurred about 25 million years ago, which is recent enough that, while past its prime, it may not be completely dead. Large plains of solidified lava cover the surface around the shield volcanoes.

Olympus Mons, photographed by ESA's *Mars Express* orbiter.

As if the biggest volcano was not enough, one of the largest canyons is also found on Mars. Vallis Marineris is a trench in Mars's equatorial area, running more than 2,500 miles long and reaching depths of four miles. If this canyon were located on Earth, it would extend from one end of the United States to the other. Vallis Marineris is likely a rift valley formed in parallel to the colony of shield volcanoes, similar to the rift valleys on Venus. But the walls of Vallis Marineris seem to have been eroded by the flow of fluids. Whether these walls were sculpted by lava, ice, water, or a combination of all three, is not known, although water is suspected to have played at least some role.

Unlike Venus, not all of Mars's history has been obliterated by lava, and its surface shows several signs that water once flowed. Since most of the northern hemisphere of Mars is relatively flat and lower in elevation than the southern hemisphere, one theory is that this was the site of an ancient ocean. The lack of obvious shorelines challenge the idea, although shoreline imprints could have been erased by huge tsunamis caused by asteroid and comet impacts.

In other areas, outflow channels are long, wide tracks of washed terrain, extending up to hundreds of miles in length. These channels were likely carved by liquid water during huge outburst floods, several billions of years ago. Valley networks, in contrast, are much smaller features, resembling dried up river basins on Earth.

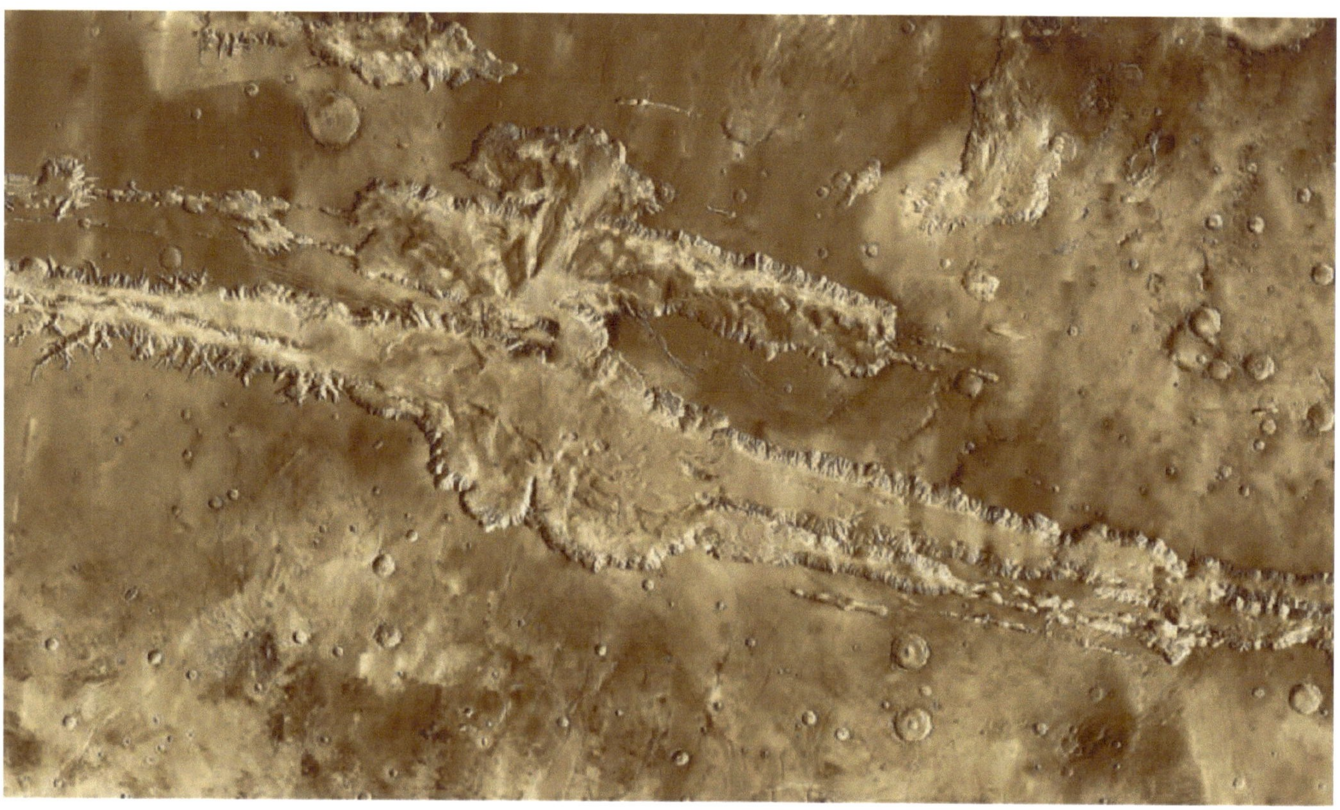

A mosaic image of Vallis Marineris by NASA's *Mars Odyssey* orbiter.

The fan-shaped remains of an ancient river delta on Mars, imaged by NASA's Mars *Global Surveyor* orbiter.

Whether or not microbes ever swam in these ancient waters, it seems inevitable that Mars is where humankind will one day step foot. Future inhabitants will surely miss the vibrant Earthly vistas. But despite its cold temperatures, dust storms, and lack of breathable air, Mars is the world most similar to our own. Our long-term survival as a species may depend upon settling it. In the 19th century, we looked at Mars and thought we saw a dying civilization. We now look at the dead planet and see our second home.

Moon of Saturn

Titan

THE WATER CYCLE ON EARTH is special. As sunlight heats Earth's seas, water evaporates and forms into clouds. Over time, water then precipitates from these clouds, which streams back into the seas, only to evaporate and start the process again. This dynamic cycle sculpts Earth's landscapes, controls regional climates, and powers all life on the planet, including human civilization. Titan, Saturn's largest moon, is the only other world in the Solar System with such a cycle. But Titan's rains, rivers, and seas are not made of water.

Discovered in 1655 by astronomer Christiaan Huygens and named after a race of giants in Greek mythology, Titan is larger than planet Mercury. Unique among all the moons of the Solar System, Titan has a thick atmosphere, with a pressure slightly greater than what we experience on Earth. Due to the high pressure and Titan's weaker gravity, we could actually fly on this world, as long as we had wings to flap.

Like our home planet, the primary component of Titan's atmosphere is nitrogen, but methane and other hydrocarbons are also present and create global layers of haze. This orange smog envelopes the moon and cloaks its surface. Why Titan has a dense atmosphere, but similarly sized icy moons Ganymede and Callisto do not, is unknown. Since sunlight should have converted the atmospheric methane to more complex hydrocarbons within 50 million years, there must be an ongoing process to replenish it.

In the 2000s, we finally landed a probe on Titan, and peered through the clouds using infrared and radar imaging. Underneath the haze, Titan is revealed as a world with polar lakes. With an average global temperature of -290°F, Titan's surface is far too cold for water to exist as a liquid. These lakes are instead made of methane and other organic compounds, a toxic stew similar to gasoline.

Artistic depiction of one of Titan's polar lakes.

The largest lake, Kraken Mare, is about the same size as the Caspian Sea, and some of the lakes on Titan are more than 300 feet deep. Mysterious, bright blobs of material floating in these alien seas may be hydrocarbon icebergs. Unlike on Earth, no waves have been observed, although there is evidence of shoreline erosion caused by wave activity. If we could stand on the shore of one of these seas, we would see beneath the tangerine clouds a mostly flat and gentle sea, perhaps with ripples, dimly lit by the distant Sun.

Most of Titan's lakes are in the northern polar region, whereas only a few lakes exist near the south pole. This asymmetry is mysterious, but may be caused by seasons. Like Saturn and the rest of its large moons, Titan's axis is tilted 27 degrees. The north pole, in the past two decades, has been in a long winter. Possibly, the smaller, shallower lakes will evaporate in the summer, and re-emerge in the southern hemisphere. Rivers of methane trickle across Titan's landscapes, carrying liquids into and around its many lakes. Dried river basins in lower latitudes tantalizingly hint at Titan's past climates. Streams and rivers on Titan sculpt networks of canyons and fjords, some more than 1,800 feet deep.

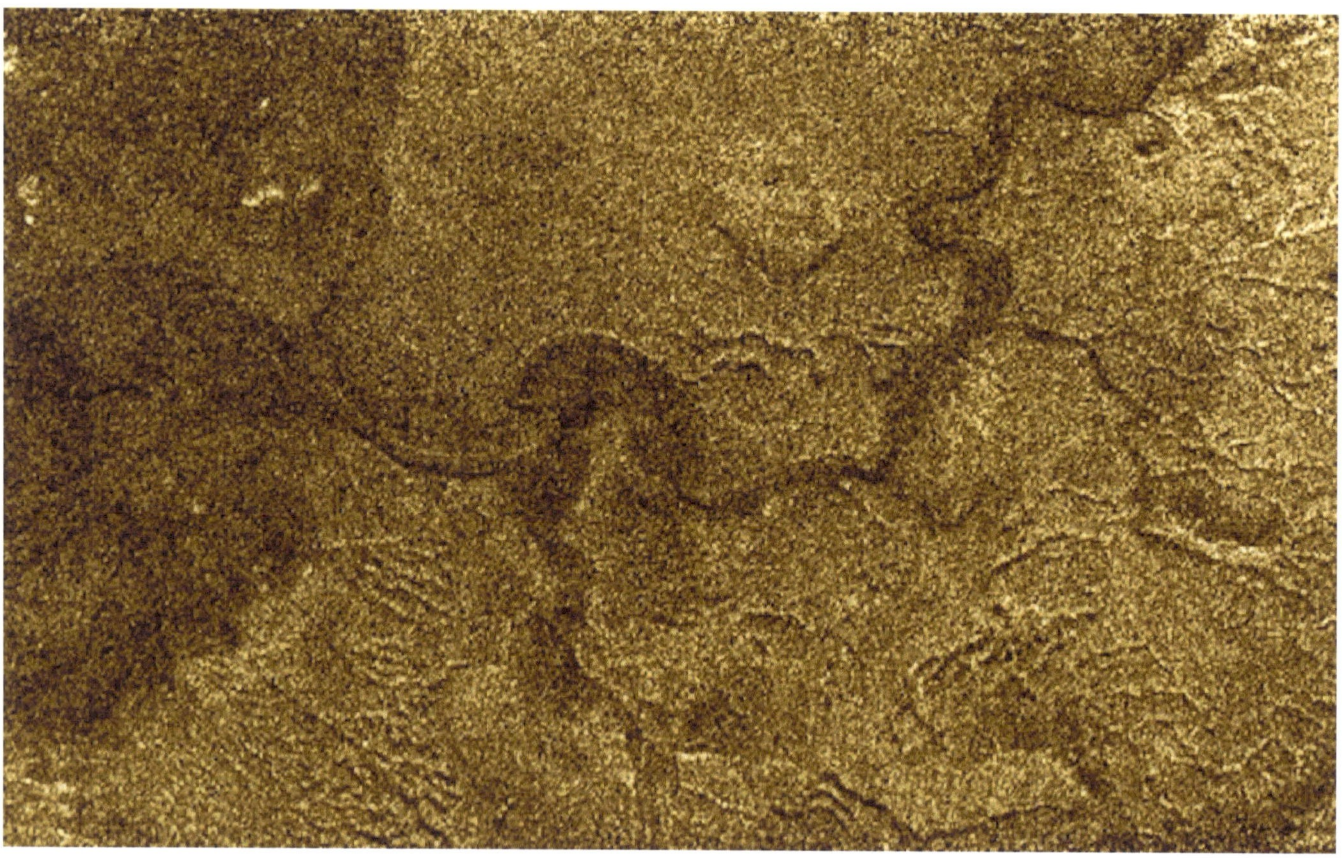

A radar image of a flowing river of methane on Titan, from NASA's *Cassini* orbiter.

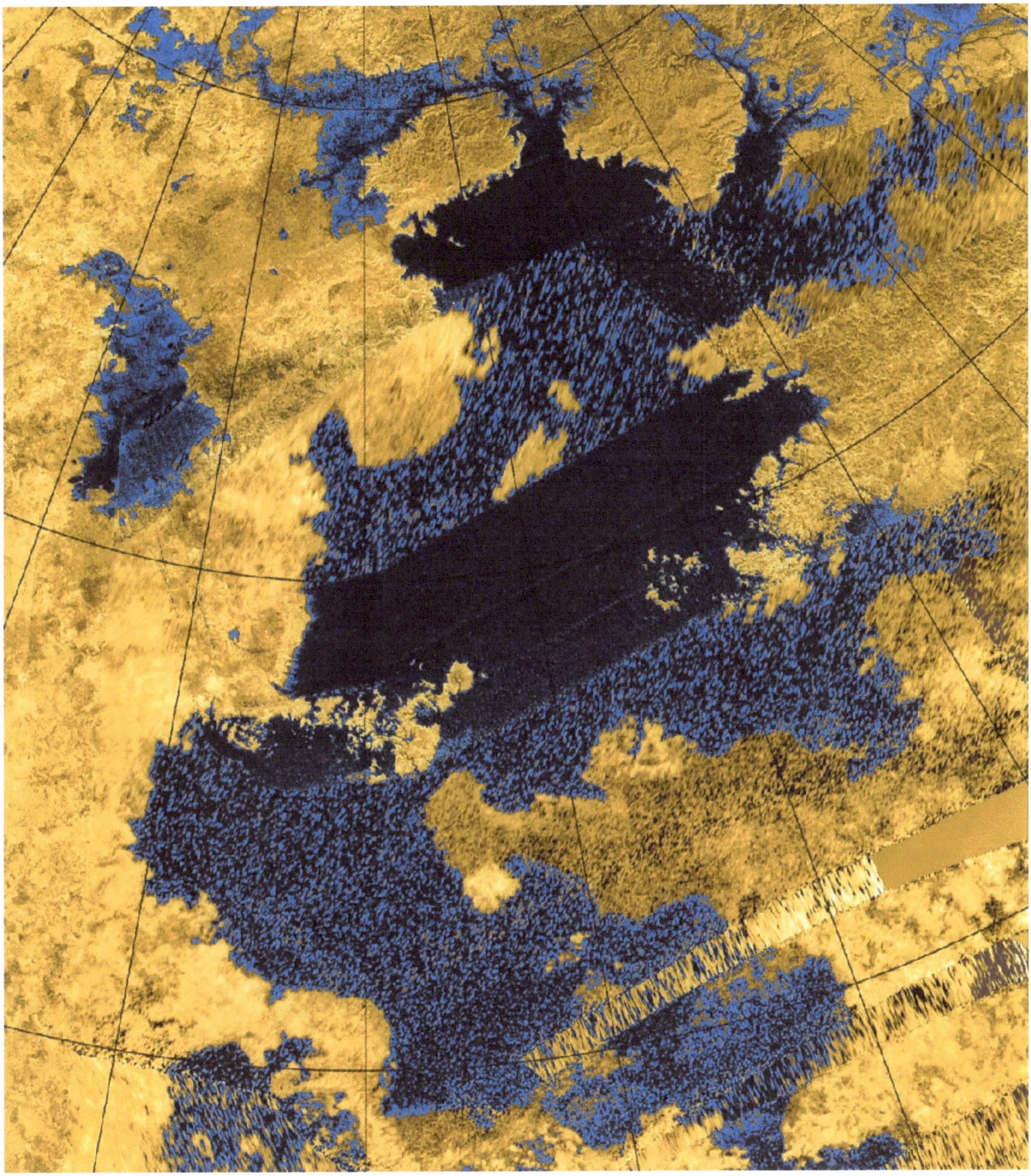

NASA's *Cassini* orbiter took the above radar image of Kraken Mare, Titan's largest lake.

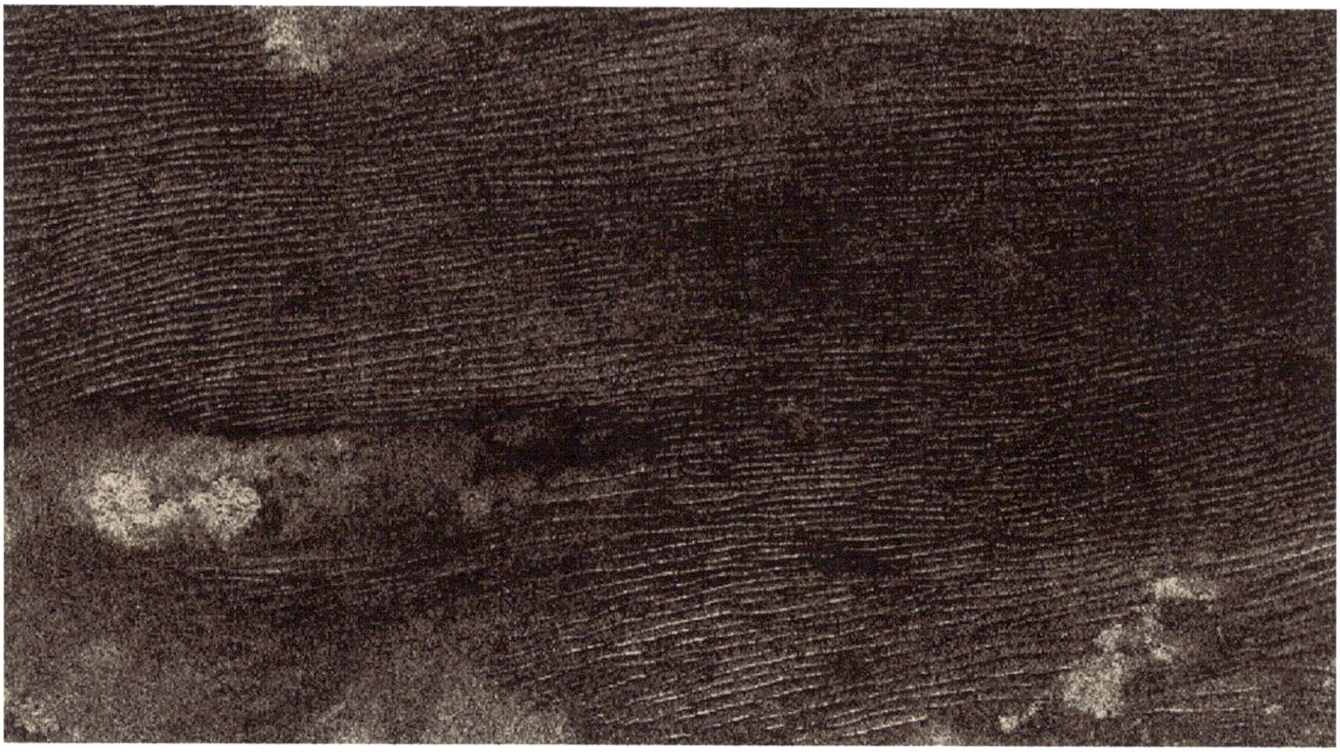

Dunes of organic sands stretch across Titan's tropics, captured in radar image from NASA's *Cassini* orbiter.

Titan's seasons affect its weather too. Studies suggest the moon has annual monsoonal downpours of more than a foot of rain per day. The most intense floods seem to occur in the mid-latitudes, where more "tropical" air meets the colder polar air masses. Other regions of Titan experience persistent morning drizzles, probably a result of clouds pushing upslope and then condensing to form rain. Interestingly, Titan is the only other world on which we could see rainbows. But on Titan, since sunlight is dim and the atmosphere is thick, its rainbows would mostly be in infrared light. Our best chance to see rainbows on Titan would be to wear night-vision goggles.

Away from the polar regions, Titan is a dry, desert world. Vast fields of dunes, up to 330 feet high, stretch across its equator. Dunes make up 20% of its surface and cover more than three times the total area of all of Earth's dunes. Unlike the dunes on Earth, which are mostly made of silicate sands, the soot in Titan's dunes is made of organic material. Curiously, Titan's dunes point to the east, the opposite direction of the general wind direction in the tropics. The sands probably erode out of the moon's bedrock during heavy rainstorms, and then are oriented to the east during rare but powerful westerly wind bursts. With all this loose sediment, giant dust storms occur on Titan just as they do on Mars.

Titan also shows signs of tectonics and volcanism. Chains of icy mountains up to two miles high may have been thrust upward by compression in Titan's crust. One isolated mountain stands next to an irregular depression about 20 miles across, which is interpreted as a complex volcanic structure. Another possible volcano has been identified, but why Titan only has a few is a mystery. Perhaps older volcanoes are buried by erosion and sediments, an echo of the disappearing volcanoes we explored on Ceres. Volcanoes on Titan, in contrast to the ones on Earth, would erupt a slushy mix of water and ammonia, with trace amounts of methane, possibly supplying the methane in its atmosphere.

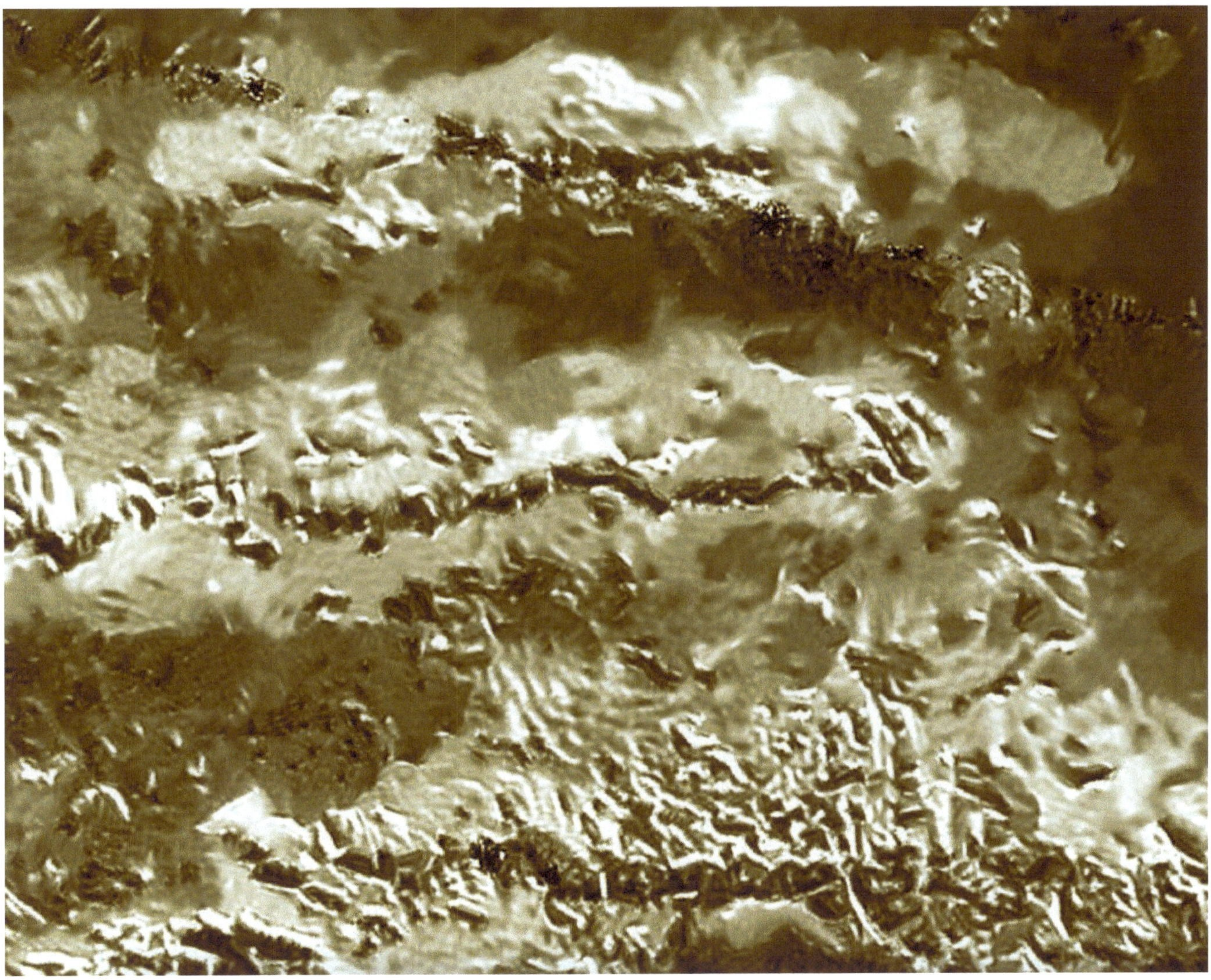

A radar image of parallel mountain belts up to two miles tall, from NASA's *Cassini* orbiter.

Sunlight reflecting off a hydrocarbon lake on Titan, from NASA's *Cassini* orbiter.

Simulated image of Sotra Patera, a volcanic construct on Titan which may spew watery lava onto the organic-covered surface.

Gravity measurements indicate Titan's icy crust is up to 50 miles thick, and sits atop a layer of liquid water and ammonia more than 100 miles deep. Due to the increasing compression in Titan's interior, this ocean itself rests upon a layer of high-pressure ice, all of which envelopes a large rocky core. It is unlikely that any of the organics on the surface seep into Titan's ocean, given its extreme depth. Titan's underground ocean, not in contact with rocky or organic material and mixed with ammonia, is expected to be lifeless. On the surface, ponds of organic stew evoke the prebiotic soups of early Earth, but the frigid temperatures and lack of liquid water are major obstacles for life to actually arise.

And yet, like a siren song, Titan calls us to its open seas. With misty air, plenty of fuel for energy, familiar scenery, and a comfortable pressure, Titan is a promising new home for humankind, after the Moon, Mars, and Callisto. Our descendants will just need proper insulation from the extreme cold, and then they will, literally, soar. Titan itself has a glorious future too. In a few billion years, when the Sun swells into a red giant star, Titan's atmosphere will warm enough that it can maintain liquid water on its surface. Perhaps then, when Earth has been burnt to a crisp, Titan will take its place as the oasis of the Solar System.

Moon of Neptune

Triton

WELCOME TO TRITON, the dazzling rebel moon of Neptune. Discovered in 1846 by astronomer William Lassell and named after the Greek sea god, Triton travels on a tipped path, significantly inclined to Neptune's equator. It is also the only large moon in the Solar System to orbit backwards, opposite to its planet's rotation. To explain such a weird orbit, this oddball must have been born circling the Sun among the planets.

For Neptune to snag a world as large as Triton, about three quarters the size of our Moon, Triton would have had to somehow slow down. Either Triton collided with one of Neptune's original moons, or it was part of a binary system, in which gravitational interactions with Neptune split the two worlds apart, catching Triton but banishing its twin. At first, Triton's orbit around Neptune was highly elliptical, tidally heating the moon through constant stretching and squeezing as it revolved farther and closer to the giant planet. Today, its orbit has circularized. But Triton, the moon, has not settled.

With a surface temperature of -395°F, Triton is one of the coldest worlds in the Solar System. Surprisingly, this icebox moon is still a dynamic, active world. Like Earth and Mars, Triton has polar ice caps that change with its seasons. But given its orbital inclination, and Neptune's own axial tilt, Triton's seasons are more extreme. As a result, its caps dramatically grow and shrink. In the winter darkness, the polar cap extends almost to the equator, but in the summer, it may completely vanish. Along the boundary of the cap are a series of dark patches, appearing like giant mushrooms peaking out of freshly fallen snow. These may be regions where so much ice has vaporized that the darker bedrock is exposed.

The polar cap of Triton, imaged by NASA's *Voyager 2* spacecraft.

Unlike the water ice caps of Earth, or the caps of frozen water and carbon dioxide on Mars, Triton's polar caps are made of nitrogen ice. At Triton's temperatures, nitrogen cycles between an ice and gas, and so the moon has a very thin atmosphere composed mostly of nitrogen. Triton's atmospheric pressure is only a fraction of even Mars's atmosphere, but it is substantial enough for nitrogen ice particles to form wispy clouds. Wind patterns transport air from warmer to cooler regions, and a blue-tinged haze envelopes the moon, caused by trace amounts of hydrocarbons. Some of these hydrocarbons sink onto the glistening surface and paint the moon in melanges of rose and lavender.

Emanating from Triton's southern polar cap are plumes of nitrogen gas and dust rising five miles above the ground. Although Triton is billions of miles away from the Sun, these plumes may be solar-powered geysers. When the amount of sunlight increases in the spring, the ice underneath the top layer warms, and enough pressure builds to force the eruptions, like larger and longer-lasting versions of Mars's geysers.

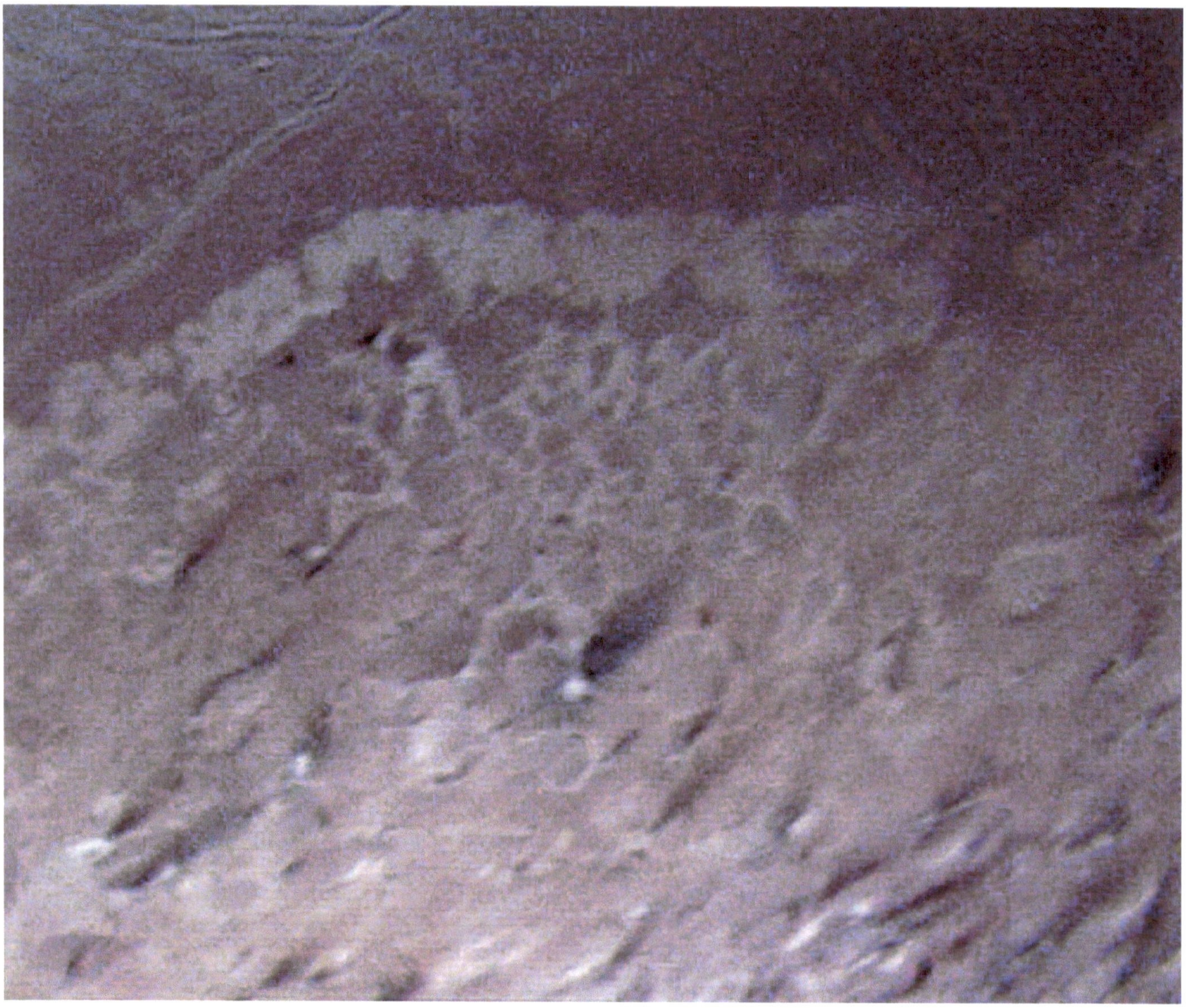

NASA's *Voyager 2* spacecraft photographs plume material blown by wind.

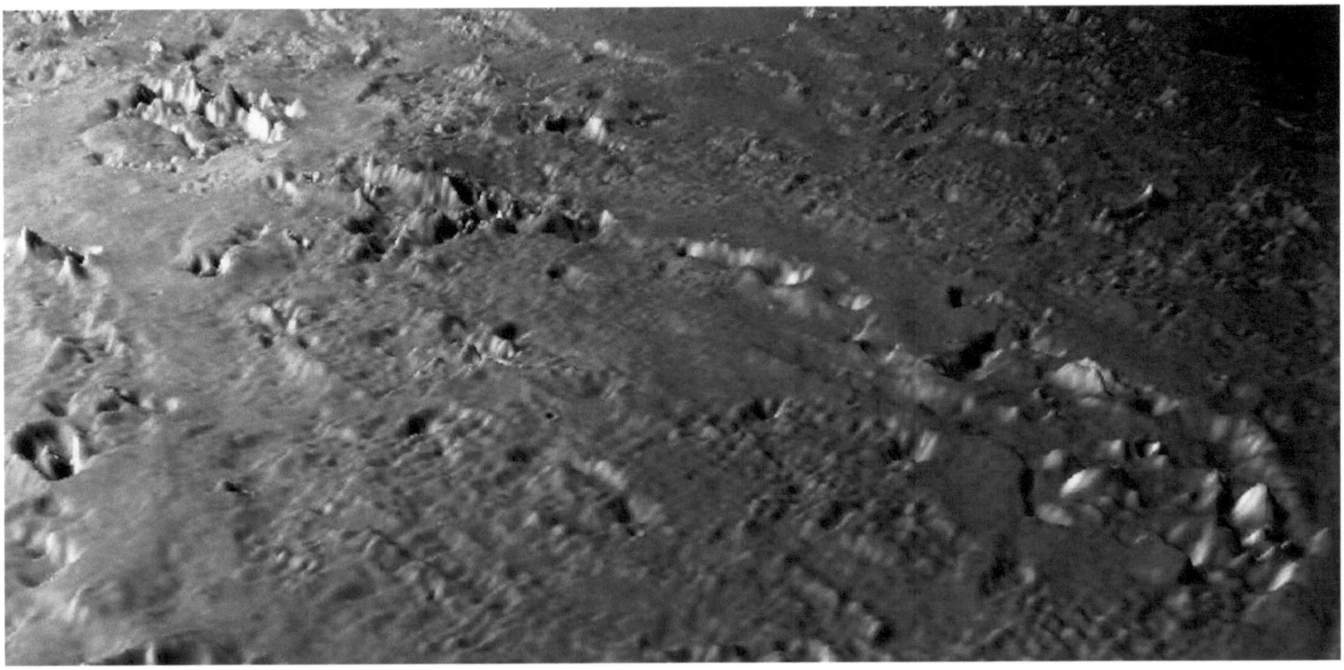

Perspective view of icy volcanic plains of Triton.

Standing on Triton's surface at a safe distance from these eruptions would offer a wonderful vista. We can imagine dark, ashy plumes gushing high into the sky from the brighter, icy ground, as the azure Neptune hangs gracefully overhead. Perhaps we would see all the plumes turning the same corner, as they are blown by Triton's upper-level winds. Wafting far into the horizon, the plume material is carried up to 100 miles away before falling back onto the surface, leaving a series of dark streaks across Triton's polar ice cap.

Triton's surface is remarkably young, with very few impact craters. Like Europa, Triton is a smooth moon, with only minimal variations in topography. The strangest geological area is the cantaloupe terrain, appropriately named because it looks exactly like the skin of a cantaloupe. The formation of these unique textures is not certain, but may involve rising of more buoyant ice through layers of denser ices. Warped tectonic ridges, up to thousands of miles long, cut through this and other terrains. Some of these are reminiscent of the double ridges of Europa, and may have formed from similar tidal stresses.

Other areas of Triton are covered in vast, smooth plains of solidified lava. Pits up to 60 miles wide are interpreted as volcanic calderas, and are surrounded by solidified lava flows. A pair of plains bound by scalloped walls appear to be frozen lava lakes. On frigid Triton, all this lava would be an exotic potion of water, ammonia, methane, nitrogen, and possibly other chemicals. Based on the low amount of impact craters, some of these flows may be less than 10 million years old, a blink of an eye on a geological scale. This suggests Triton's volcanoes are still active, and its icy lavas will continue to flood its surface.

Since Triton's orbit is no longer eccentric, it does not experience tides in the manner of Io, Europa, and Enceladus. However, Triton may still be heated by tides from its orbital inclination. Radioactive decay in the rocky core, inferred from its high density, is also expected to warm Triton's interior. These sources of heat should maintain an ocean of water and ammonia, buried under more than 100 miles of rigid ice. However, due to the noxious concentration of ammonia that would be required to keep it liquid, any ocean underneath Triton's crust is likely to be uninhabited.

Triton shares a lot in common with Titan. Both are the largest moons of their parent planets. Both have nitrogen atmospheres and organics on their icy surfaces. Both are likely to have subsurface water-ammonia oceans. But while Titan will blossom in the distant future, Triton is cursed to a cataclysmic end. Triton is slowly falling toward Neptune, and in about three billion years, it will cross the Roche limit, the point at which moons cannot exist because the planet's tidal forces are too strong. When this happens, Triton will shatter into billions of pieces and become a brilliant ring of Neptune.

For now, Triton's vigor in the coldest reaches of the Solar System reminds us that in nature, we should always expect the unexpected. Keep this in mind as we journey to the next, and last, world on this tour.

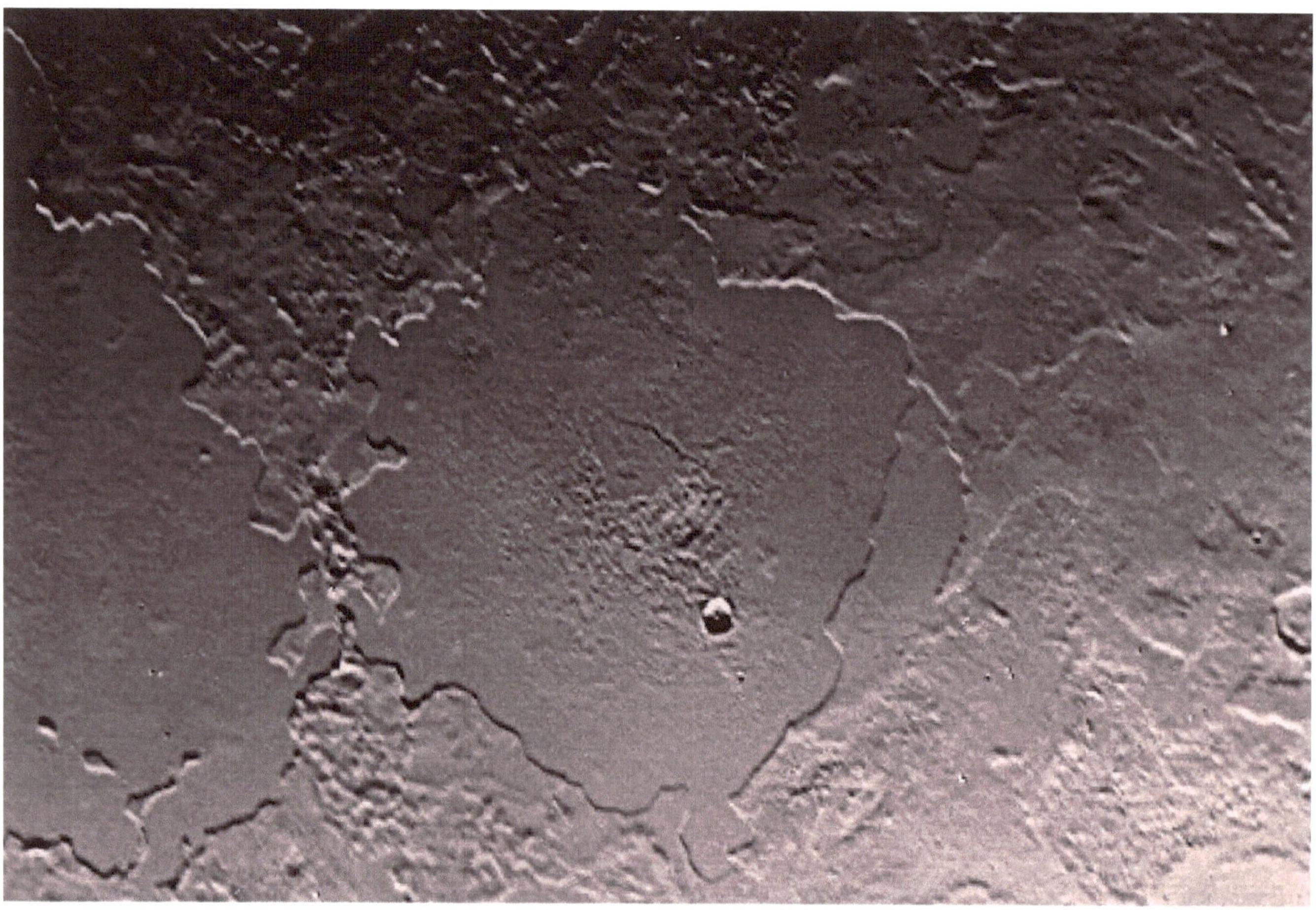

Frozen lakes of lava, seen by NASA's *Voyager 2* spacecraft.

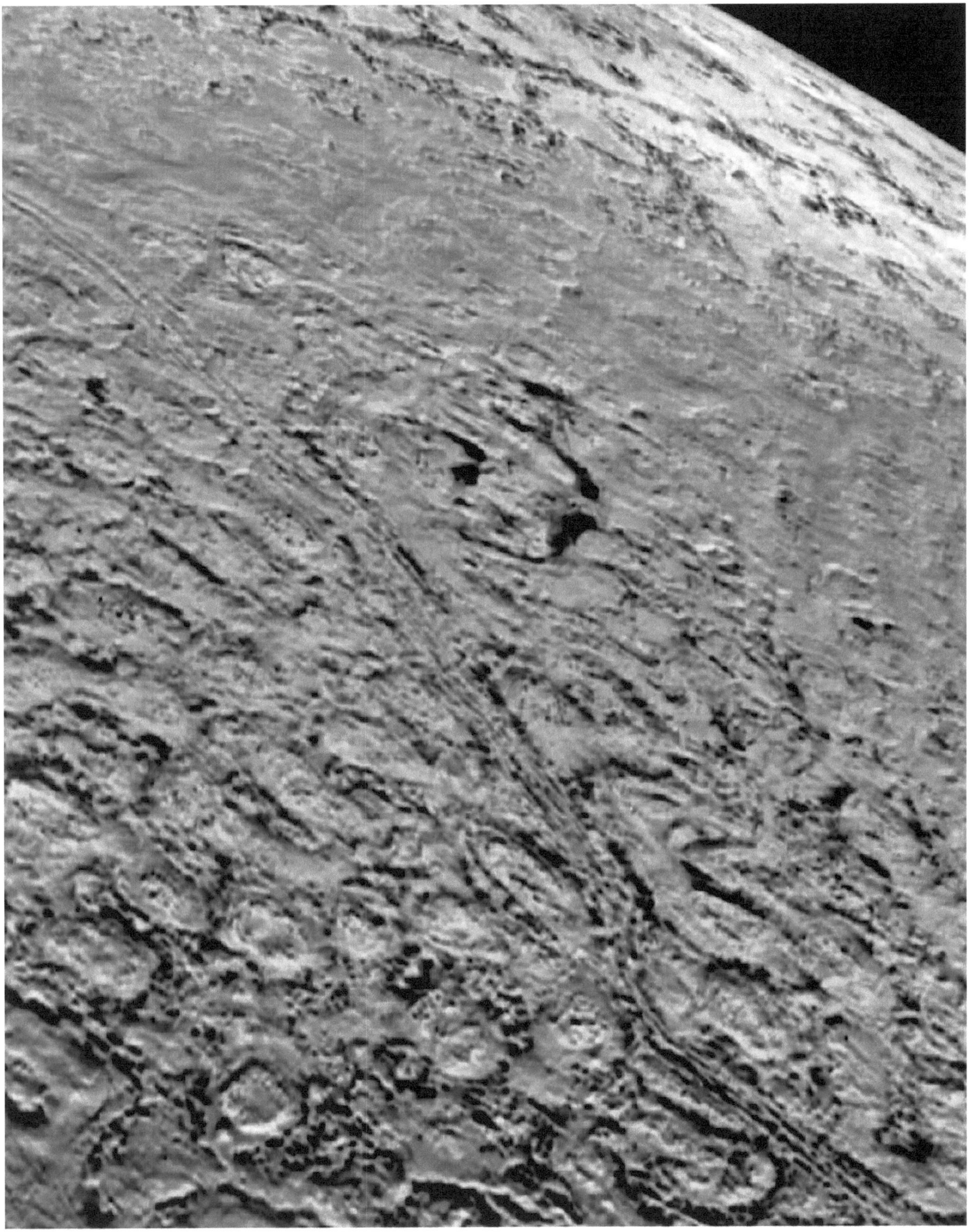

Complex lands, probably formed by a combination of processes. Image from NASA's *Voyager 2* spacecraft.

Ice Dwarf

Pluto

WHEN ASTRONOMER CLYDE TOMBAUGH discovered Pluto in 1930, it was hailed as the ninth planet. But Pluto always stood out. Not obeying the motions of the other planets, the icy world orbits the Sun on an inclined tilt. Pluto is also smaller, about half the width of Mercury and 4% its mass. Beginning in the 1990s, more objects with similar eccentric orbits were discovered in Pluto's realm. Many are lumpy chunks of ice, but at least a handful are large enough to be round worlds, and one is almost the same size as Pluto. This disc of ice dwarfs is named for astronomer Gerard Kuiper who hypothesized its existence in the 1950s. Pluto, rather than the runt of the planets, is the king of the Kuiper Belt.

Far from being a frozen relic, and despite temperatures never climbing above -370°F, Pluto is a dynamic, complex wonderland. Like Triton, Pluto has an atmosphere composed mostly of nitrogen, with small amounts of methane and other hydrocarbons. This atmosphere, much thinner than that of Mars, nonetheless displays multiple layers of blue hazes, wind patterns, and probably low-level wispy clouds. Because Pluto is tipped on its side and has a highly elliptical orbit which brings it even closer to the Sun than Neptune, its seasons are wild. When Pluto is farthest from the Sun, its atmosphere is expected to completely freeze and fall to the surface.

Pluto has one of the most varied surfaces in the Solar System. The dichotomy between dark and bright regions is nearly as extreme as on Iapetus, with the dark regions confined mostly to its "tropics" and the brighter ices draped over the polar latitudes. The glaring exception is Pluto's heart. Etched in the middle of the ice dwarf, this cosmic valentine tattoo is the size of the Gulf of Mexico. Upon closer view, the heart is made of individual polygonal segments, separated by narrow, shallow indentations as if they were all puzzle pieces carefully linked together.

Pluto's atmosphere backlit by the Sun, from NASA's *New Horizons* spacecraft.

Closeup of Pluto's heart as seen by NASA's *New Horizons* spacecraft.

The shapes of these polygons suggest Pluto's heart is a giant sea of convecting nitrogen ice. Convection, the movement of lighter material upward and denser material downward, cycles blobs of nitrogen ice from the surface to a few miles below, and then back to the surface, with each blob creating the distinct polygons. Given that there are no impact craters on this feature, it must be continually replenished within just hundreds of thousands of years.

Feeding this sea of convecting ice are nitrogen glaciers, flowing into the heart as solid rivers. In the chaffed landscapes just outside the heart, concave imprints and curved margins hint at past glaciers, perhaps when Pluto's climate was different. All this flowing nitrogen ice cycles to and from the ice dwarf's atmosphere as its winds and seasons change, just as Earth's weather and seasons control its water cycle. Pluto also has polar ice caps, but they are thinner than Triton's and made mostly of methane frost. Like nitrogen, methane on Pluto changes between a gas and ice. The interplay of these two volatiles, freezing and vaporizing in complex and overlapping cycles, makes Pluto one of the Solar System's most chemically rich worlds.

Bordering Pluto's heart are a series of majestic mountain chains, with peaks rising up to four miles above the ground, comparable to the Alps. The Plutonian mountains are sharp, indicating they are young and may even still be rising. Unlike the mountains of Earth, the ragged peaks on this world are not made of rocks. At Pluto's cold temperatures, its bedrock, and the composition of its mountains, is solid water ice. Several mountains are capped with methane frost, reminiscent of the snowcapped peaks on Earth. Pluto's mountains may be shards of crust which were uplifted by flowing glaciers of nitrogen ice, a much different process than the tectonic forces which give rise to Earth's mountains. Even more strangely, some of the water ice mountains on Pluto appear to be drifting within the heart, like icebergs in a frozen sea.

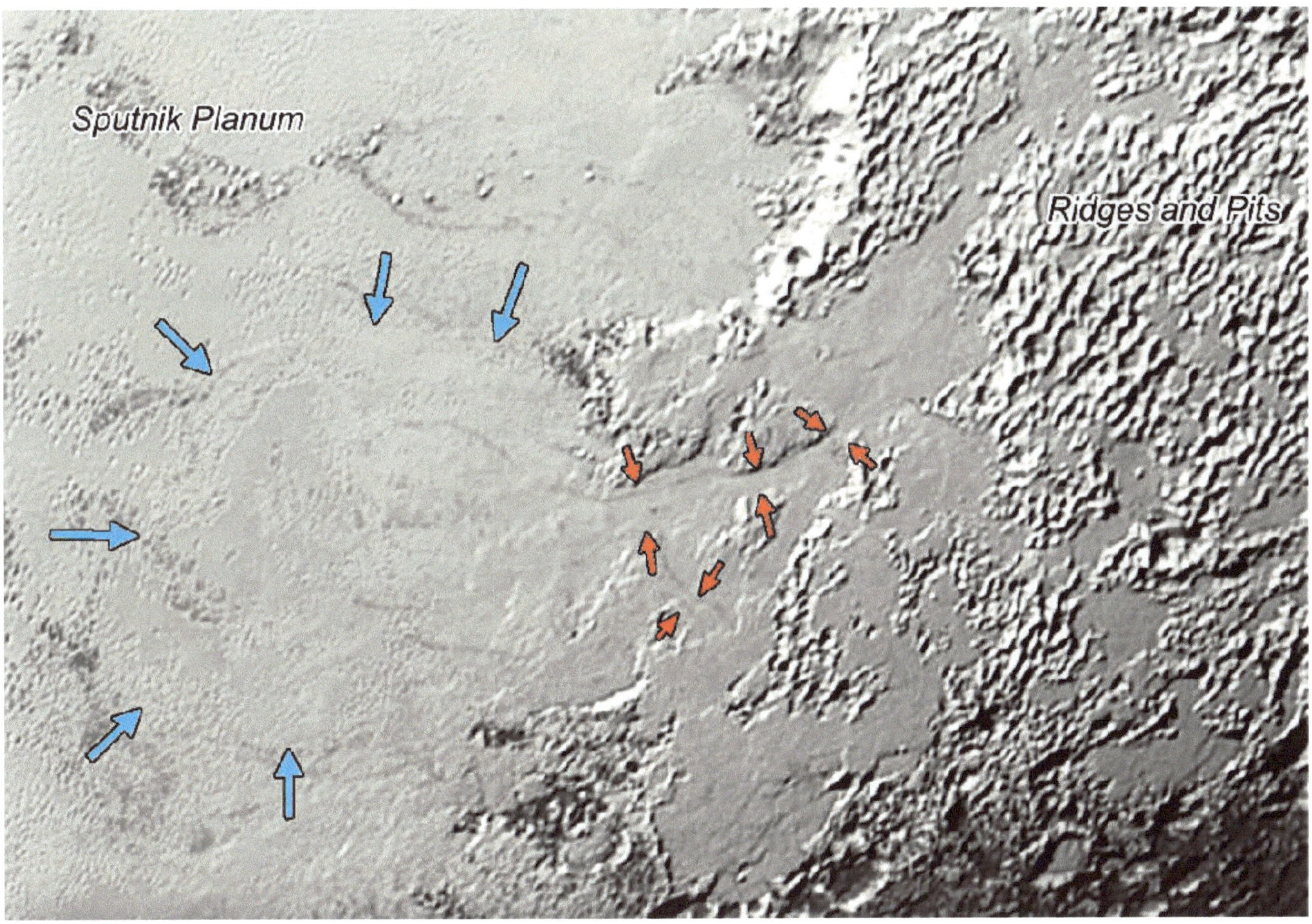

NASA's *New Horizons* spacecraft captures flowing glaciers of nitrogen ice. Blue arrows indicate ice flowing within the flat plains, whereas the red arrows point to valley glaciers.

One picture of Pluto from NASA's *New Horizons* spacecraft says so much: On the left, mountains of water ice tower four miles above the ground.

Surprisingly, Pluto's atmosphere is thick enough to sculpt sand dunes. Vast areas of parallel dunes ripple across its heart. Given that the heart of Pluto is young and active, these dunes must be forming today. The key to understanding how frigid Pluto could have shifting sands is in its mountains. As winds blow across Pluto's peaks, small bits of methane frost are carried onto the heart, where they sublimate, turning from ice to a gas. In this process, ice particles are also lifted and then blown by breezes to create the dunes.

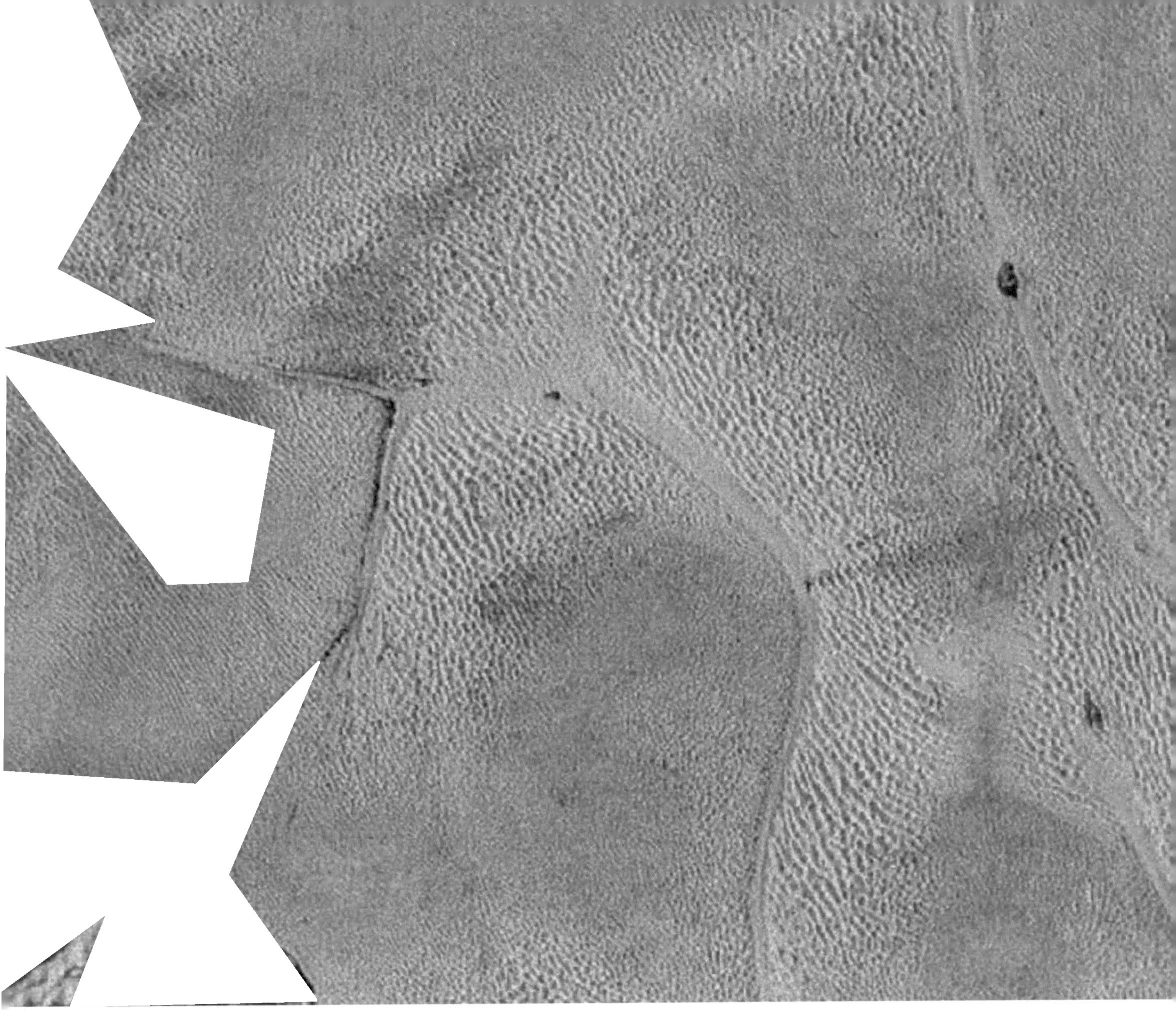

On the right side of this picture, convecting cells of nitrogen ice are overlayed by rippling dunes of frozen methane grains.

Walking in areas of Pluto's heart would offer some of the most spectacular scenery in the Solar System. We would perhaps step through gentle, rolling dunes, our boots kicking up glittering sands of methane ice. Looking in one direction, the horizon would be flat. Behind us, colossal mountains with frosted slopes would loom tall. This landscape of Pluto never faces its moon Charon, so the sky would be dim and starry, with a layered blue haze draped over the horizon. The only sound in this magical wintery twilight would be total silence.

The vaporization of ices plays other roles on Pluto. East of the heart are lands pierced by blades of methane ice as tall as skyscrapers. Because these blades are at a high elevation, they are thought to form when methane gas deposits into ice and is stacked as towers. As Pluto's climate changes, ice sublimates along the margins, leaving behind the remarkable bladed texture. A similar process occurs in the high elevations of the Chilean Andes, which creates ice structures called penitentes. But Pluto's penitentes are far more magnificent than the ones on Earth, which are only a couple feet high.

Other regions of Pluto are equally perplexing. In Pluto's older, darker terrains, a collection of craters are crowned by bright, ghostly halos, thought to be deposits of methane ice. Narrow canyon networks, reaching depths of two miles, slice through Pluto. Some of the canyon walls are painted in a bright frost, probably methane which has deposited in the shadows.

In Pluto's southern hemisphere, two giant domes are roughly 100 miles across and up to four miles high, with wide central depressions. The domes are comprised of multiple small mounds. These unique structures are thought to either be large volcanoes, or a colony of multiple volcanic vents which erupt a thick, goopy substance and merged to create the domes. Lava on Pluto would be made of water and ammonia, possibly mixed with other substances. The lack of craters indicates icy lava has erupted in these regions within the last billion years.

The gigantic methane blades of Pluto, appearing like snakeskin, seen by NASA's *New Horizons* spacecraft.

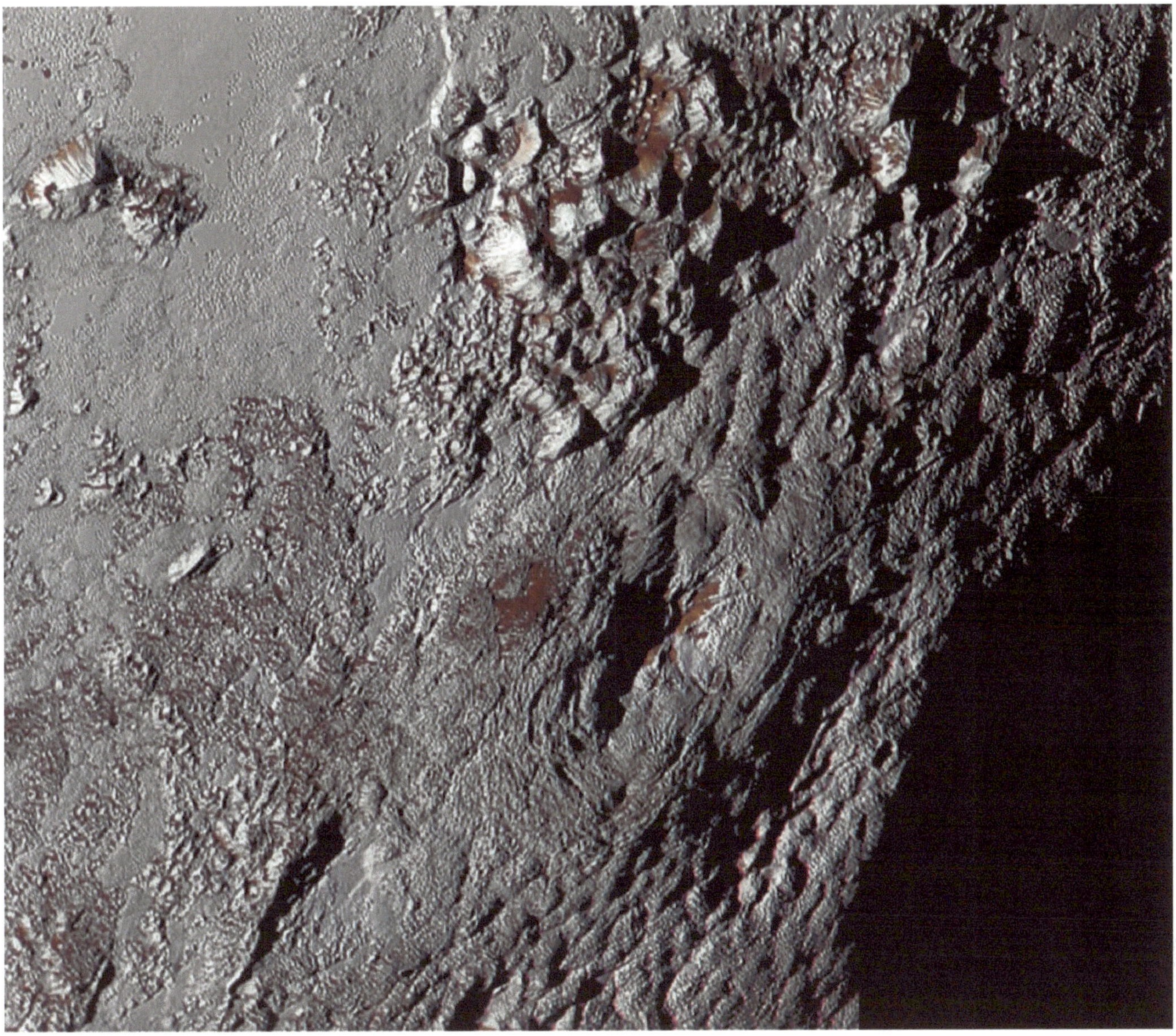

Wright Mons, possibly a giant ice volcano, as imaged by NASA's *New Horizons* spacecraft.

There are signs that Pluto has an ocean of water and ammonia, more than 100 miles below the surface. Such an ocean could be maintained by radioactive decay in Pluto's rocky core, and is probably in the process of freezing. We should not expect to find alien microbes swimming in these cold, murky depths. But if Io showed how tidal heating can power volcanoes on cold moons, Pluto heralds an equally profound truth: small, faraway places can sustain active geology without warmth from tides. Imagine, then, what wonders await us in the future, on even more distant worlds.

BEYOND

NONE OF THE PHOTOGRAPHS we have viewed would have been possible without the scientists who had the vision and dedication to send robotic space probes to these distant lands. From the *Voyager* probes, to *Galileo* and *Cassini*, to *New Horizons* and *Juno*, to *Messenger* and *Dawn*, and to all the rovers on Mars, these spacecraft have revolutionized our understanding of the cosmos. While our journey in this book comes to a close, the journey itself is not over. The Solar System offers even more worlds worth exploring.

The four largest moons of Uranus have barely been studied, and call for a proper expedition. One of these moons, Ariel, has deep canyons and solidified flows of icy lava on their floors, suggesting this world may still be active. Titania, the largest moon of Uranus, is more heavily cratered with long, bright-walled canyons. Umbriel is mysteriously darker than the other large moons of Uranus, but a white donut-shaped deposit hints at more wonders. On Oberon, the farthest of Uranus's large moons, dark material has oddly pooled in some of the craters.

Background artwork: Sedna at the far edge of the Solar System.

Several mysterious worlds beyond Pluto have not been imaged as anything more than points of light. Haumea is particularly strange: rotating every four hours, it is one of the fastest spinning objects in the Solar System. This swift rotation flattens Haumea into the shape of a football. Another large ice dwarf is Makemake, which has an abundance of methane and may experience some of the same surface processes as Pluto. Eris, the most massive world beyond Neptune and only slightly smaller than Pluto, is also one of the brightest places in the Solar System, hinting at active geology. Quaoar and Gonggong are reddish, smaller ice dwarfs, but still massive enough to have experienced past tectonics and volcanism. And Sedna, a similarly sized red object, is on such an extreme orbit that it takes more than 11,000 years to circle the Sun.

The more we learn about all the worlds in the Solar System, the more we truly understand our own planet, and our place in the universe. The robotic spacecraft we send to these destinations are the pioneers, scouting and examining the lands that perhaps our descendants will one day walk. As long we dream big and aim high, our journey to other worlds has just begun.

IMAGE CREDITS

CC BY 2.0: *https://creativecommons.org/licenses/by/2.0*
CC BY 3.0: *https://creativecommons.org/licenses/by/3.0*
CC BY-SA 3.0 IGO: *https://creativecommons.org/licenses/by-sa/3.0/igo*
Page 1: NASA, color edited by author
Page 4-5: NASA/JPL-Caltech/MSSS/Texas A&M University
Page 6: NASA/JPL-Caltech
Page 7: NASA/Goddard Space Flight Center
Page 8: NASA/Goddard Space Flight Center/Alan Friedman CC BY 2.0 *https://flickr.com/photos/24662369@N07/7556375618*
Page 9: NASA/JPL-Caltech
Pages 10-11: NASA/JPL-Caltech/Space Science Institute
Page 12: NASA/JPL-Caltech/SwRI/MSSS/Kevin M Gill CC BY *https://photojournal.jpl.nasa.gov/catalog/PIA23801*
Page 13: NASA/JPL-Caltech/SwRI/MSSS/Kevin M. Gill CC BY 2.0 *https://flickr.com/photos/kevinmgill/49803032983*
Page 14: NASA/JPL-Caltech/SwRI/MSSS/Navaneeth Krishnan S CC BY 3.0 *https://jpl.nasa.gov/images/pia25692-nasas-juno-views-northern-cyclones-on-jupiter*
Page 15: NASA/JPL/Space Science Institute
Page 16: NASA/JPL-Caltech/SSI/Cornell
Page 17: NASA/JPL-Caltech/SSI/Kevin M. Gill CC BY 2.0 *https://flickr.com/photos/kevinmgill/27723918159*
Page 18: *NASA/JPL-Caltech/Space Science Institute, colorized by author*
Page 19: NASA/JPL-Caltech/SSI/Kevin M. Gill CC BY 2.0, cropped *https://flickr.com/photos/kevinmgill/32788158595/in/album-72157688015492704/*
Page 20: NASA/JPL-Caltech/SSI/CICLOPS/Kevin M. Gill CC BY 2.0 *https://flickr.com/photos/kevinmgill/44350267281*
Page 21: NASA/JPL-Caltech/Space Science Institute
Page 22: NASA/JPL-Caltech
Page 23: NASA/JPL
Page 24: NASA/JPL
Page 25: NASA, ESA, CSA, STScI
Page 26: NASA/JPL, color edited by author
Page 27: NASA/JPL, color edited by author
Page 28: NASA/JPL
Page 29: NASA/JPL, color edited by author
Pages 30-31: NASA/JPL
Page 32: NASA/Johns Hopkins University Applied Physics Laboratory/Arizona State University/Carnegie Institution of Washington
Page 33: NASA/Johns Hopkins University Applied Physics Laboratory/Carnegie Institution of Washington/National Astronomy and Ionosphere Center, Arecibo Observatory
Page 34: NASA/Johns Hopkins University Applied Physics Laboratory/Carnegie Institution of Washington, colorized by author
Page 35: NASA/Johns Hopkins University Applied Physics Laboratory/Carnegie Institution of Washington, colorized by author
Page 36: NASA/GSFC/Arizona State University
Page 37: Illustration by Lucian Rudaux
Page 38: NASA/GSFC/Arizona State University
Page 39: NASA/JPL
Page 40: NASA/JPL-Caltech/UCLA/MPS/DLR/IDA
Page 41: NASA/JPL-Caltech/UCLA/MPS/DLR/IDA/ASI/INAF
Page 42: NASA/JPL-Caltech/UCLA/MPS/DLR/IDA, colorized by author
Page 43: NASA/JPL-Caltech/UCLA/MPS/DLR/IDA/PSI
Page 44: NASA/JPL-Caltech/Kevin M. Gill CC BY 2.0 *https://flickr.com/photos/kevinmgill/38926064465*
Page 45: NASA/JPL, colorized by author
Page 46: NASA/JPL, colorized by author
Page 47: NASA/JPL
Pages 48-49: NASA/JPL/Space Science Institute
Page 50: NASA/JPL-Caltech/SwRI/MSSS/Kevin M. Gill CC BY 2.0 *https://flickr.com/photos/kevinmgill/51238659798*
Page 51: NASA/JPL/Brown University, colorized by author
Page 52: NASA/JPL
Page 53: Diagram by Wikipedia user kelvinsong, cropped *https://commons.wikimedia.org/wiki/File:Ganymede_diagram.svg*
Page 54: NASA/JPL/Space Science Institute
Page 55: NASA/JPL/Space Science Institute, colorized by author
Page 56, left: NASA/JPL/Space Science Institute
Page 57: NASA/JPL-Caltech/Space Science Institute, colorized by author
Page 58: NASA/JPL-Caltech, colorized by author
Page 59: NASA/JPL, colorized by author
Page 60: NASA/JPL, colorized by author
Page 61: NASA/JPL, colorized by author
Page 62: NASA/Johns Hopkins University Applied Physics Laboratory/Southwest Research Institute
Page 63: NASA/Johns Hopkins University Applied Physics Laboratory/Southwest Research Institute
Page 64: NASA/Johns Hopkins University Applied Physics Laboratory/Southwest Research Institute

Page 65: NASA/Johns Hopkins University Applied Physics Laboratory/Southwest Research Institute
Pages 66-67: NASA/Johns Hopkins University
Page 68: NASA/JPL/University of Arizona
Page 69: NASA/JPL/University of Arizona
Page 70: NASA/JPL/USGS
Page 71: NASA/JPL-Caltech/Kevin M. Gill CC BY 2.0 *https://flickr.com/photos/kevinmgill/50098173893*
Page 72: NASA/JPL/USGS
Page 73: NASA/JHUAPL
Page 74: NASA/JPL-Caltech/SwRI/MSSS/Kevin M. Gill CC BY 3.0 *https://photojournal.jpl.nasa.gov/catalog/PIA25695*
Page 75: NASA/JPL-Caltech/SETI Institute
Page 76: NASA/JPL-Caltech
Page 77: NASA/JPL/University of Arizona
Page 78: NASA/JPL/University of Arizona/University of Colorado
Page 79: NASA/JPL-Caltech/SETI Institute
Page 80: NASA/JPL-Caltech/Space Science Institute
Page 81: NASA/JPL/Space Science Institute
Pages 82-83: NASA/JPL/Space Science Institute
Page 84: NASA/JPL/Space Science Institute
Page 85: NASA/JPL-Caltech
Pages 86-87: NASA/JPL-Caltech
Page 88: NASA/JPL-Caltech
Page 89: NASA/JPL
Page 90: NASA/JPL
Page 91: NASA/JPL
Page 92: NASA/JPL
Page 93: JAXA/ISAS/DARTS/Kevin M. Gill CC BY 2.0 *https://flickr.com/photos/kevinmgill/50517022233/in/album-72157716561343202/*
Page 94: NASA/JPL-Caltech/USGS
Page 95: NASA/JPL/MSSS
Page 96: NASA/JPL-Caltech/University of Arizona, colorized by author
Page 97, top: ESA/DLR/FU Berlin CC BY-SA 3.0 IGO *https://dlr.de/en/latest/news/2021/04/20211221_an-icy-spring-at-the-martian-south-pole*
Page 97, bottom: ESA/DLR/FU Berlin CC BY-SA 3.0 IGO *https://esa.int/ESA_Multimedia/Images/2018/12/Perspective_view_of_Korolev_crater*
Page 98: NASA/JPL-Caltech/MSSS
Page 99: ESA/DLR/FUBerlin/Andrea Luck CC BY 2.0 *https://flickr.com/photos/192271236@N03/52770375578*
Page 100: NASA/JPL-Caltech/Arizona State University
Page 101: NASA/JPL/Malin Space Science Systems
Page 102: NASA/JPL-Caltech/SSI/Kevin M. Gill CC BY 2.0 *https://flickr.com/photos/kevinmgill/40047599334*
Page 103: NASA/JPL-Caltech
Page 104: NASA/JPL-Caltech/ASI, colorized by author
Page 105: NASA/JPL-Caltech/ASI/USGS
Page 106: NASA/JPL-Caltech/ASI, colorized by author
Page 107: NASA/JPL-Caltech/ASI, colorized by author
Page 108: NASA/JPL/University of Arizona/DLR
Page 109: NASA/JPL-Caltech/ASI/USGS/University of Arizona, color edited by author
Page 110: NASA/JPL
Page 111: NASA/JPL
Page 112: NASA/JPL
Page 113: NASA/JPL/Universities Space Research Association/Lunar & Planetary Institute, colorized by author
Page 114: NASA/JPL, colorized by author
Page 115: NASA/JPL, colorized by author
Page 116: NASA/Johns Hopkins University Applied Physics Laboratory/Southwest Research Institute/Alex Parker
Page 117: NASA/JHUAPL/SwRI
Page 118: NASA/Johns Hopkins University Applied Physics Laboratory/Southwest Research Institute
Page 119: NASA/Johns Hopkins University Applied Physics Laboratory/Southwest Research Institute, colorized by author
Pages 120-121: NASA/Johns Hopkins University Applied Physics Laboratory/Southwest Research Institute
Page 122: NASA/Johns Hopkins University Applied Physics Laboratory/Southwest Research Institute
Page 123: NASA/Johns Hopkins University Applied Physics Laboratory/Southwest Research Institute
Pages 124-125: NASA/JPL-Caltech

www.ingramcontent.com/pod-product-compliance
Lightning Source LLC
LaVergne TN
LVHW071241210926
848486LV00007B/317